BRINGING THE NCTM STANDARDS TO LIFE

EXEMPLARY PRACTICES FOR MIDDLE SCHOOLS

Yvelyne Germain-McCarthy

D1307622

EYE ON EDUCATION
6 DEPOT WAY WEST, SUITE 106
LARCHMONT, NY 10538
(914) 833–0551
(914) 833–0761 fax
www.eyeoneducation.com

Library of Congress Cataloging-in-Publication Data

Germain-McCarthy, Yvelyne, 1948–
    Bringing the NCTM standards to life : exemplary practices for middle schools / by Yvelyne Germain-McCarthy.
        p. cm.
    Includes bibliographical references.
    ISBN 1-930556-13-6
    1. Mathematics—Study and teaching (Middle school)—United States. I. Title.

QA13.G472 2001
510′.71′273—dc21

                                                                        00-067717

10 9 8 7 6 5 4 3 2

Editorial and production services provided by
Richard H. Adin Freelance Editorial Services
52 Oakwood Blvd., Poughkeepsie, NY 12603-4112
(845-471-3566)

Also Available from EYE ON EDUCATION

Bringing The NCTM Standards to Life:
Exemplary Practices From High Schools
by Yvelyne Germain-McCarthy

Assessment in Middle and High School Mathematics:
A Teacher's Guide
by Daniel J. Brahier

A Collection of Performance Tasks and Rubrics:
Middle School Mathematics
by Charlotte Danielson

A Collection of Performance Tasks and Rubrics:
High School Mathematics
by Charlotte Danielson and Elizabeth Marquez

Teaching Mathematics in the Block
by Susan Gilkey and Carla Hunt

Open-Ended Questions in Elementary Mathematics:
Instruction and Assessment
by Mary Kay Dyer and Christine Moynihan

A Collection of Performance Tasks and Rubrics:
Upper Elementary School Mathematics
by Charlotte Danielson

A Collection of Performance Tasks and Rubrics:
Primary School Mathematics
by Charlotte Danielson and Pia Hansen

Mathematics the Write Way:
Activities for Every Elementary Classroom
by Marilyn S. Neil

ACKNOWLEDGMENTS

I'd like to express my sincere appreciation to the teachers profiled in this book for their hard work and commitment to improving the teaching and learning of our children. In particular, I wish to thank them for sharing their ideas and responding to my numerous communications.

Acknowledgment is due to Robert Sickles, Kathryn Bremner, Jamie Graber, Judy Keeley, Gilbert Cuevas, and Deborah Gober for suggestions that sparked noticeable improvements in the book. My students and colleagues have also contributed valuable insights through conversations, reflections and readings of sections of the books.

My consultant affiliation with Southwest Educational Development Laboratory (SEDL) has been a helpful venture. I wish to thank Steven Marble for his leadership in a number of SEDL projects that have served to heighten my awareness of the challenges inherent in the interactions between research, theory and practice, within school settings. SEDL colleagues that have helped to enrich this process include Glenda Copeland, Chris Ferguson, Sandra Finley, Como Molina, and Maria Torres.

I offer special thanks to my family: my parents, Georges and Eugenie Germain, for their love and unfailing support; my brothers, Gerard Germain, Serge Germain, and Claude Germain, and my dear friend Emilio Delrio for their love and confidence.

Most importantly, I extend loving gratitude to my husband, Henry McCarthy, my older son, Julian McCarthy, and my younger son, Germain McCarthy, for their love, patience, and encouragement.

Finally, I wish to dedicate this book to children in impoverished sections of the world where freedom, or the means to participate in the types of learning environments described in this book, are but a dream. I donate my proceeds from this book to promoting the best education available to children from Riviere Froide, a small town in my homeland, Haiti.

FOREWORD

The National Council of Teachers of Mathematics document, *Principles and Standards for School Mathematics*, stresses the importance of providing every child with rich opportunities to learn high-quality mathematics. An important part of achieving this goal is helping competent teachers engage students in active learning environments wherein students develop a deep understanding of mathematical skills, concepts, and relationships.

Bringing the *NCTM Standards to Life: Exemplary Practices for Middle Schools* shows readers how classroom teachers may successfully implement many of the goals of NCTM's Standards. This volume profiles teachers who strive to meet the NCTM vision of a high-quality mathematics education for every child by reflecting on their own practice or collaborating with others. The reader is invited into teachers' classrooms to witness a dynamic environment where students are engaged in thinking about mathematics and challenging each other as mathematicians. Readers get access to teachers' conversations about classroom instruction and gain insight into how to make the vision of the Council a reality.

Noteworthy is the identification of teacher successes in addressing issues related to teaching mainstreamed students, students in special education, and students in low- performing school settings. Lesson overviews that summarize key ideas from each profiled classroom facilitate the reader's use or adaptation of lessons. A special feature of this work is the author's commentary on connecting lessons to broader issues in mathematics education research. This book is a valuable companion to *Principles and Standards for School Mathematics* because it illustrates a variety of good teaching practices that teachers can implement.

Lee Stiff
North Carolina State University

REFLECTIONS FROM THE FIELD

Our American students have fallen behind their counterparts in other countries, according to the Third International Mathematics and Science Study. Although the National Council of Teachers of Mathematics (NCTM) has developed Standards to be implemented in schools throughout the United States, only a small, trickling effect is making its way into local school districts. Simply having these Standards developed into Mathematics curricula isn't enough. An important piece is to provide professional development for our teachers to help them facilitate students' entry and participation in the world of mathematics.

Yes, our teenagers are thirsty for knowledge, even though some may refuse to show it. They all need classrooms that are dynamic and involving, frustrating and exhausting, invigorating and inviting. Lessons, which put them in the driver's seat, need to be developed and implemented.

Bringing the NCTM Standards to Life: Exemplary Practices for Middle School Teachers by Yvelyne Germain-McCarthy profiles teachers and teacher groups who have successfully created such classroom environments for students. The lessons demonstrated in *Exemplary Practices* show how the use of active learning, technology, and community relations work towards making that happen. These are not the average, everyday, "Okay students, let's turn to page 74 and work problems 1 to 30" type of lessons. While they call for a deeper understanding to take place, they are also exciting and challenging for the students. Readers will find it worthwhile to take these lessons back to the mathematics classroom to enhance their students appreciation and understanding of mathematics.

Allison Gregerson
Roosevelt Middle School
Kenner, Louisiana

TABLE OF CONTENTS

ABOUT THE AUTHOR

Yvelyne Germain-McCarthy is an associate professor of mathematics education at the University of New Orleans, where she teaches the elementary and secondary mathematics methods courses for graduate and undergraduate students. She received her B.S. in mathematics and M.Ed. in mathematics education from Brooklyn College. She earned her Ph.D. in mathematics education from Teachers College at Columbia University. She taught high school and middle grades mathematics for 17 years. As co-project director of systemic initiative grants in Louisiana, she leads reform efforts to guide a collaborative of universities in the greater New Orleans area toward implementing reformed-based instruction in classes of pre-service mathematics and science teachers. She is a frequent speaker at professional conferences and, as a member of the National Faculty, serves as a consultant to school districts. She is the author of Bringing the *NCTM Standards to Life: Exemplary Practices from High Schools.*

PREFACE

More than a decade has passed since the release of the *Curriculum Standards for School Mathematics*, by the National Council of Teachers of Mathematics (NCTM). Are teachers implementing these standards today, and if so, how? Larry Cuban, an education professor at Stanford University said, "The NCTM *Standards* have certainly shaped how people talk about mathematics. There has been tremendous action in terms of districts and states adopting them. When you talk about implementation, you get much more variation. That is not unlike what happens with every single curriculum reform in the past century." Both he and past NCTM President Glenda Lappan agree that "teachers have been highly selective in what portions of the standards they rely on. A very tiny percentage have adopted them as a whole" (Hoff 2000).

PURPOSE OF THIS BOOK

This is a book for anyone interested in gaining insight into how the reform movement in mathematics, as advocated by NCTM, is being implemented by teachers in the United States. Teacher educators will find the book useful for exemplifying NCTM reformed-based strategies for preservice and in-service teachers, as well as provide an answer to a question many of my students ask: "Where are the real teachers who are doing this stuff, and how do they do it?" For courses that combine both high school and middle grades mathematics teachers, my companion book from Eye on Education, *Bringing the NCTM Standards to Life: Exemplary Practices from High School Teachers*, plays a similar role.

The teacher profiles that compose the heart of the book are descriptions constructed from classroom visits, written statements, interviews, and videotapes of how the teachers implement *standards*-based lessons in their classrooms. Nine profiles of teacher educators across the nation who have gone beyond mere awareness of *Curriculum Standards* to conceptualizing and implementing new curricula that conform to NCTM *Principles and Standards* for the middle grades are highlighted.

Chapter 1 presents overviews of the NCTM *Standards* documents and some of the research that provided the rationale for its constructivist framework. Chapter 2 describes key elements of exemplary practices and a description of NCTM's vision. Chapters 3 and 4 are profiles of teachers who are successful at implementing *Principles and Standards* with students at different ends of the achievement spectrum or with students having limited English proficiency. Chapters 5–8 profile teachers teaching innovative curriculum to mainstream students. Chapter 9, written for a high school general mathematics class, has interesting uses of the Internet directly applicable to middle grades students. Chapter 10 shows how collaborations between teachers, or university mathematics and education professors collaborating with teachers, can be powerful forces in creating engaging activities for students. Chapter 11 profiles a school-university-community effort to implement the Japanese approach to curriculum and teaching to improve students' learning in a school that was taken over by its state due to low performance on high-stakes tests. In each chapter, I include a Dialogue between Colleagues section, where we clarify or expand ideas from the profile. The Unit Overview summarizes key ideas for each profile, and a final Commentary highlights the specific standards, issues, or research that informed the strategies the teachers used.

Although the Unit Overviews specify grade levels, readers will find that they can be easily modified. Ideas for extensions of the curricula will emerge not only because of the richness of the activities but also because the lessons move from the concrete to the abstract. This reform-based approach provides ways to vary the emphasis of the concepts presented. Finally, the summary chapter reflects on the vision behind the NCTM *Standards* and on what this vision means for students and the mathematics community, now and in the future.

The profiles may incorporate a number of different content standards and they all reflect the NCTM principles for equity, curriculum, teaching, learning and assessment. Most demonstrate classroom applications of all of the NCTM process standards: problem solving, reasoning, connection, communication, and representation. The profiles will give readers ideas about how to implement the standards and an opportunity to learn what may be a new mathematical concept or tool. Figure P.1 summarizes the standards addressed in each profile. Areas that are highlighted in the table represent major foci in a profile. Join me in learning from these exemplary teachers!

FIGURE P.1 PRINCIPLES AND STANDARDS: APPLICATIONS IN THE PROFILES

Chapter: Teacher(s)	Principles	Content	Process
3: Susan Mercer & Tammra Detviler	1-6 Equity	Numbers, Measurement, Geometry, Algebra	1-5 Problem Solving Communication Representation
4: Laura Mullen	1-6 Assessment	Numbers, Measurement, Geometry, Algebra	1-5 Problem Solving Reasoning Representation
5: Madeline Landrum	1-6	Numbers, Measurement, Geometry, Algebra	1-5 Problem Solving Reasoning
6: Kim McReynolds	1-5 Assessment	Numbers, Measurement, Geometry, Algebra	1-5 Problem Solving Communication Connection Representation
7: Merrie Schroeder	1-6	Numbers, Geometry, Data Analysis, Measurement	1-5 Communication Connection Representation
8: Kim Leblanc & DarleneMorris	1-5	Algebra, Geometry, Measurement	1-5 Problem Solving Communication
9: Jim Specht	1-6 Technology	Numbers, Measurement, Geometry, Algebra	1-5 Problem Solving Connection Representation
10: KatherineOwens, Richard Sanders, & Robert Lipsinki	1-5 Assessment	Numbers & Relations	4-3 Problem Solving Communication
11. Public School 2	1-6 Curriculum	Numbers, Geometry, Measurement	1-5 Problem Solving Communication

1

THE NCTM
STANDARDS

What "Principles and Standards for School Mathematics" has done is to set higher standards for our students and for ourselves. The goal of both the revised and past documents, then and now, is to provide a guide for sustained efforts to improve students' mathematics education.

Lee Stiff,
President, National Council of
Teachers of Mathematics

In this quote from the "President's Message" column of the May/June (2000a) *News Bulletin* of NCTM, Stiff succinctly summarizes the major focus of NCTM's reform efforts. Past NCTM President Glenda Lappan further highlights the key components: "What is the reform of mathematics teaching and learning guided by NCTM's Standards all about? My answer is that we are about the following three things: upgrading the curriculum, improving classroom instruction, and assessing students' progress to support the ongoing mathematics learning of each student" (Lappan 1998).

In 1989, NCTM recommended that we teach and assess students in very nontraditional ways. Today, these goals are still important, and others have been updated to depict the vision and directions for school mathematics programs described in NCTM's *Principles and Standards for School Mathematics* (hereafter called *Principles and Standards*). The overall purpose of the *Principles and Standards* is to revise and clarify the unique trilogy of NCTM standards published in 1989, 1991, and 1995 (hereafter called the *Standards documents*), which defined standards for content, teaching, learning, and assessment of K-12 mathematics programs.

In her message, Lappan further noted that two commitments inform NCTM's reform efforts: inclusiveness and understanding. All students should experience effective mathematics teaching, and the focus of mathematics instruction should be to help students develop a deep understanding of important mathematics concepts (3). Support of both commitments requires that teachers believe that all students can learn challenging mathematics, that they know how to probe current understandings of students, and that they can present students with engaging tasks that help them connect new knowledge to old knowledge.

The theory of constructivism is useful in efforts to implement these commitments because it is practiced or experienced in an environment in which learners are trying to make sense of a problematic situation in order to understand an idea. Constructivism is the framework for NCTM's reform efforts.

CONSTRUCTIVISM

In the pedagogical philosophy known as constructivism, learners make sense of experiences by constructing their own knowledge about the world around them. Jean Piaget (1973) wrote,

> To understand is to discover [that]…a student who achieves a certain knowledge through free investigation and spontaneous effort will later be able to retain it: he will have acquired a methodology that can serve him for the rest of his life, which will stimulate his curiosity without the risk of exhausting it. At the very least, instead of his having his memory take priority over his reasoning power…he will learn to make his reason function by himself and will build his ideas freely….The goal of intellectual education is not to know how to repeat or retain ready-made truths. It is in learning to master the truth by oneself at the risk of losing a lot of time and of going through all the roundabout ways that are inherent in real activity (106).

Goldin (1990) adds, "for large number of students at all levels of mathematics education, methods involving the statement and application of rules…are less successful than methods involving of mathematical discovery (i.e., methods based on a *constructive* learning model)" (46). Thus, learning is not simply the acquisition of information and skills; it also includes the acquisition of a deep understanding that requires time but that enables the learner to better construct meaning from a problem. Learning occurs when a novel situation contradicts the learner's beliefs and therefore requires new constructs to make sense of and interpret the situation (Confrey 1990).

Simon (1995) notes that constructivism does not define a specific way to teach mathematics. Rather, it "describes knowledge development whether or not there is a teacher present or teaching is going on" (117). However, this does not imply that teachers have little impact on the learning of their students. Some constructivists, such as von Glasersfeld (1993), believe teachers can orient students in certain directions for building knowledge. Planning for teaching is considered a teacher's responsibility, and plans must be revised and modified as needed during the learning process. Attention to the misconceptions of students provides teachers with a rich source of information that allows them to detect and understand in which areas students need guidance. The teacher's task is to help students learn to find tools that are useful for solving problems—ideally, problems that students have identified from their own work.

Some constructivists view the small group process, in which students work together on mathematical tasks that require a high level of communication about a problem, as a crucial component of the development of conceptual understanding. Social interaction as an essential factor in a learner's organization of experiences underlies the theory of social constructivism. According to Vygotsky (1978), "Any function in the child's cultural development appears twice on two planes. First it appears on the social plane, and then on the psychological plane" (57). A focus on the processes by which a learner constructs meaning from social interactions opens a window for researchers to examine those processes because it "constitutes a crucial source of opportunities to learn mathematics in that the process of constructing mathematical knowledge involves cognitive conflict, reflection, and active cognitive organization....As such, mathematical learning is...an interactive as well as [a] constructive activity" (Cobb, Wood, and Yackel 1990, 127).

Although constructivism focuses on teaching students to generate their own understanding, it should not be confused with the "discovery learning" reforms that were advocated for both mathematics and science in the 1960s. According to Fensham, Gunstone, and White (1994), the major difference between the two approaches is that constructivist strategies are "learner centered but teacher controlled in a way [such] that there is always something that the learner is called upon to construct" (6). Thus, in the inquiry- or discovery-based curricula of the 1960s, it was assumed that students learned a concept or idea when they discovered it. In contrast, the constructivist approach begins with what the students already think and believe regarding a particular idea. Students' attempts to verify these ideas then serve as catalyst for the learning process.

SKILLS FOR CITIZENS OF THE TWENTY-FIRST CENTURY

Is it important to a student's future to engage in mathematical activities that extend beyond the memorization of procedures? In their book about educating for the twenty-first century, Caine and Caine (1997) describe why traditional ways of thinking about teaching and learning need to change:

The amount of information being created and made available almost exceeds the imagination. Moreover, the information is overwhelming traditional channels. In time, these traditional channels will be unable to cope with the flood; and unlike most floods, the information flow will not dissipate....There is no way education can be in charge of information....As a result, traditional sources of information for students are fundamentally inappropriate. Irrespective of whether textbooks are effective, few textbooks can be the primary source of important and current information. Similarly, even teachers who are constantly updating their own professional expertise can only keep pace with a small fragment of what is becoming available (46).

Thus, locating, managing, sharing, and using information to help solve problems are the basic skills of the future that students should practice in all subjects. To elucidate the nature of the mathematics that should be studied as well as the constructivist learning environment described by the *Principles and Standards,* I will briefly review the current work of brain-based researchers to show how closely it supports NCTM's *Standards* documents.

BRAIN-BASED RESEARCH

Brain-based researchers view the mind's design as that of a "pattern detector." Learners continually search for meaning by creating patterns. Lectures or rote memorization produce a type of learning that is classified as *surface knowledge.* Although this is important, success in the twenty-first century will require *meaningful knowledge,* or knowledge that makes sense to the learner. Teaching that strives to maximize the way that the brain processes information not only enhances and increases the likelihood of meaningful knowledge but also helps citizens develop the attributes they need to thrive in the twenty-first century. Here are the attributes Caine and Caine (1997) describe (97–98, adapted slightly):

1. An inner appreciation of interconnectedness. In a world where everything is relationship, more is needed than to intellectually understand the concept of relationship. People need to have...an inner sense of connectedness that culminates in insight.

2. A strong identity and sense of being. In a fluid and turbulent world, it is very easy to become confused and disoriented. People need a coherent set of purposes, values, and beliefs. Moreover, those values should include an appreciation of life, opportunity, and respect for individual and cultural differences.

3. The capacity to flow and deal with paradox and uncertainty. We need to have ways of thinking and interpreting that helps us see

patterns in paradox. At the same time, we need to appreciate the constant mystery and to understand that at some levels, no fixed answers are possible.

4. A sufficiently large vision and imagination to see how specifics relate to each other. There is always more than we can know, and the extent of our ignorance is increasing. People frequently, naturally, and inevitably come face-to-face with the unfamiliar, the unexpected and the unknown. People therefore need an internal frame of reference that enables them to...see patterns in chaos, and to perceive commonalities.

5. A capacity to build community and live in a relationship with others. We have to be able to function both as individuals and as parts of greater social wholes.

If we accept these as desirable attributes, then schools need to develop or enhance curriculum to support the ability of students to attain these attributes in *every* subject. Describing the kind of environment and curriculum that is conducive to generating such self-directed learners has been the current focus of movements and studies that advocate reforms in the content and teaching of K-12 in-service and preservice teachers.

NCTM has for years advocated changes toward a more focused, rigorous curriculum as well as an integrated approach to teaching and assessment practices. Helping students develop the understanding and habits of mind necessary to thinking critically to address the challenges of the twenty-first century is the primary goal. The habits of mind are reflected in what Dewey (1929) calls a "disciplined mind":

> A disciplined mind takes delight in the problematic....The scientific attitude may almost be defined as that which is capable of enjoying the doubtful (228).

Qualities of people who have such a mind include "a willingness to play with ideas, to explore alternative paths or procedures, to approach situations inquisitively, to persevere, to emphasize sense making and reasoning, and to find excitement in learning" (Chapin 1997, 6). Such thinking includes the processes of science: making comparisons, generalizing, recording observations, and revising one's views because of new information.

Most students today shy away from a situation that is "doubtful." Indeed, they too often become frustrated and are quick to abandon problems for which they have no clear method of approach. According to Wheatley (1991),

> If you look at any of the work on creativity and learning, or if you look at your own creative process, it is not a nice orderly step-by-step pro-

cess that moves you towards a great idea. You get incredibly frustrated, you feel you'll never solve it, you walk away from it, and then Eureka!—an idea comes forth. You can't get truly transforming ideas anywhere in life unless you walk through that period of chaos (3).

I would add that most students who are not encouraged or allowed to pursue such experiences will likely have little desire to continue the study of mathematics and may be incorrectly perceived by some parents or teachers as not having a "math brain." What students need is a balanced curriculum in which mathematical content is learned and applied in a problem-solving environment that allows them to process and practice these attributes. Unfortunately, a 1996 report of the Third International Mathematics and Science Study, (TIMSS), the largest and most comprehensive international study ever conducted in mathematics and science education, concluded that no U.S. lesson examined for its content does so.

THE THIRD INTERNATIONAL MATHEMATICS AND SCIENCE STUDY (TIMSS)

In 1995, TIMSS gathered data on half a million students from 41 countries, focusing on student achievement, curricula, and teaching. Its eighth-grade curriculum study shows the U.S. curriculum to be more broad, more repetitive, and less challenging than those of many other countries. Videotapes of eighth-grade mathematics classes in Germany, Japan, and the United States suggest that our lessons lack emphasis on thinking and reasoning about mathematics. This may partly explain why the achievement scores of our eighth graders are low even though they spend more class time on science and math, get more homework assignments, and spend about the same amount of time watching TV as their Japanese counterparts. Results show that our fourth-grade students scored above the international average in data representation, analysis, and probability; geometry; whole numbers; fractions and proportionality; and patterns, relations, and functions. Our junior-high students, however, lagged behind most of their counterparts in other countries in math, and ranked about average in science. U.S. twelfth-grade students scored significantly below average in both mathematics and in science; the mathematics score was lower than the science score. Some educators and members of the business community express concern that our students are not adequately prepared to meet the challenges of the future. What is particularly significant about the TIMSS, however, is that the data of that study also yield insights about what needs to be improved, and the report gives suggestions about how to go about making these improvements. Have we made any gains since, based on the TIMSS data? This is one of many questions that is answered in a repeat of TIMSS study (TIMSS-R). In 1999, 38 countries agreed to assess their eighth graders, as well as their math and science programs, in an international context. Our 1995 fourth graders performed above the international average. A question of performance interest is: Four years later, how will a comparable group of eighth graders perform? Our American 13-year-olds performed better than average on the math and science skills tested in TIMSS—but they showed no improvement since TIMSS. Using the international average of 17 nations whose students participated in both studies, the results show our eighth graders performing lower in 1999 than did our fourth graders 4 years earlier. Black eighth-graders were the nation's only large ethnic group to significantly raise their math scores on TIMSS-R. However, their scores remained below the international average. TIMSS-R also conducted studies focused on highlighting the context for understanding mathematics and science teaching at the classroom level. One of the important findings is that while an average of 71% percent of students in other countries were taught by teachers having a bachelor's or master's degree in mathematics, only 41 percent of U.S. eighth graders had mathematics teachers so qualified. For further information on TIMSS and TIMSS-R, call (202) 502-7421, or click http://nces.ed. gov/pubsearch/pubsinfo.asp?pubid=2001028, or visit the Mid-Atlantic Eisenhower Consortium Web site at http://www.rbs.org/eisenhower. The Consortium disseminates information, analyses and opinions about TIMSS. To subscribe, send an e-mail message with a blank subject line to: listserv@list.rbs.org. In the body of the message, write, subscribe timss-forum.

NCTM's LEADERSHIP

The NCTM *Standard* documents arising from NCTM's ambitious vision for ways to change how mathematics is viewed and taught provide guidelines for creating a balanced curriculum, where mathematics is more than computational or algorithmic proficiency. Recognizing that information is changing and increasing at a rapid pace, the writing group for the *Curriculum Standards* wisely planned to revisit and revise it in the future. Since 1997, a *Standards 2000* group that consisted of classroom teachers, teacher educators, mathematicians, and researchers worked on the revision, which was released in April 2000 and named *The Principles and Standards for School Mathematics*. A paper copy and CD-ROM of *Principles and Standards* may be purchased by calling (800) 253-7966 or going to http://www.nctm.org/standards/ buyonline. An electronic and searchable version (dubbed *E-Standards*) with same material as the CD-ROM is available at http://www.nctm.org/.

Given that we have yet to implement much of what is in the *Curriculum Standards*, NCTM past President Gail Burrill (1996) posed the following questions to illustrate the need for a second edition:

Do history textbooks stop at the end of World War I? How does a map of Africa today compare to one from 20 years ago? What happened to science textbooks when DNA was discovered? Mathematics is no different—changes around us make changes in how we think about mathematics. Changes in what we know about how students learn affect the way we think about teaching (3).

To alleviate concerns about yet another set of major changes, she clarifies that the *Principles and Standards* document builds on and extends the foundations of the original standards publications.

PRINCIPLES AND STANDARDS FOR SCHOOL MATHEMATICS

The Writing Group sharpened and focused the *Curriculum Standards* by integrating ideas about curriculum, teaching, and assessment into one document instead of three. *Principles and Standards* consists of six principles and 10 standards that describe characteristics of quality instructional programs and valued goals for students' mathematical knowledge. Together they form the basis for developing effective mathematics instruction within four grade-band chapters: pre-K through grade 2, grades 3–5, grades 6–8, and grades 9–12. At the high school level, "All students are expected to study mathematics each of the four years that they are enrolled in high school, whether they plan to pursue the further study of mathematics, to enter the work force, or to pursue other post-secondary education"(288). A description of the particular principles and standards follow.

THE PRINCIPLES FOR SCHOOL MATHEMATICS

The *Principles and Standards* build on the solid foundation provided in the NCTM Standards Documents through a set of six principles that address the question: What are the characteristics of mathematics instructional programs that will provide all students with high-quality mathematics education experiences across the grades? Six characteristics, called guiding principles, are offered as basic tenets on which to establish quality programs and guide decisions about mathematics instruction at all levels. They focus on:

- *Equity*: Excellence in mathematics education requires equity—high expectations and strong support for all students (12).

- *Curriculum*: A curriculum is more than a collection of activities: it must be coherent, focused on important mathematics, and well articulated across the grades (14).

♦ *Teaching*: Effective mathematics teaching requires understanding what students know and need to learn and then challenging and supporting them to learn (16).

♦ *Learning*: Students must learn mathematics with understanding, actively building new knowledge from experience and prior knowledge (20).

♦ *Assessment*: Assessment should support the learning of important mathematics and furnish useful information to both teachers and students (22).

♦ *Technology*: Technology is essential in teaching and learning mathematics; it influences the mathematics that is taught and enhances students' learning (24).

THE CONTENT STANDARDS

Ten standards address the question: What mathematical content and processes should students know and be able to do as they progress through school? Of the ten, five are mathematical content standards that describe what students should know and be able to do within the areas of number and operations, algebra, geometry, measurement, data analysis, and probability and statistics. For grades 6–8, among the expectations are that students "learn significant amounts of algebra and geometry and see them as interconnected….Students are expected to learn serious, substantive mathematics with an emphasis on thoughtful engagement and meaningful learning"(212–213).

in the area of numbers and operations, students need to understand the meaning of operations with integers, fractions, decimals, and percents and how they relate to each another; use computational tools and strategies fluently; and estimate appropriately (214). In algebra, they need to understand various types of patterns and functional relationships, and use symbolic forms to represent and analyze mathematical relationships in problems or the real world (222). In geometry students should to be able to analyze the characteristics and properties of two- and three-dimensional geometric objects; select and use different representational systems, including coordinate geometry and graph theory; recognize the usefulness of transformations and symmetry in analyzing mathematical situations; and use visualization and spatial reasoning to solve problems within and outside of mathematics. "The goal is for students to systematically study geometric shapes and structure to increasingly use formal reasoning and proof in their study" (232). The study of measurement supports all students' understanding of the attributes, units, and systems of measurement and the application of these to a variety of techniques, tools, and formulas for determining measurements in the other content areas such as algebra, geometry, or science (240). In the area of analysis and probability, students should learn to formulate researchable questions; design, carry out, and interpret a survey, study, or experiment; collect relevant data and use it to make a decision; measure how confident they are in their decision; and communicate results. They should experience selecting and using appropriate statistical methods to analyze their data (248).

THE PROCESS STANDARDS

The five process standards address students' acquisition, growth in, and use of mathematical knowledge in the areas of problem solving, reasoning, connections, communication, and representation. Problem solving is the process of applying previously acquired knowledge to new and unfamiliar situations and is the primary reason for studying mathematics. In this area, all students build new mathematical knowledge through their work with problems; develop a disposition to formulate, represent, abstract, and generalize in situations within and outside mathematics; apply a wide variety of strategies to solve problems and adapt the strategies to new situations; and monitor and reflect on their mathematical thinking in solving problems (256). Recognizing reasoning and proof as essential and powerful parts of mathematics is one of the goals of this standard. Students should also make and investigate mathematical conjectures; develop and evaluate mathematical arguments and proofs; and select and use various types of reasoning and methods of proof as appropriate (262). The Communication Standard recommends that students organize and consolidate their mathematical thinking to communicate with others; express mathematical ideas coherently and clearly to peers, teachers, and others; and extend their mathematical knowledge by considering the thinking and strategies of others. They should also use the language of mathematics as a precise means of mathematical expression (268). The Connections Standard recommends that students be given opportunities to recognize and make connections among different mathematical ideas; understand how mathematical ideas build on one another to produce a coherent whole; and recognize, use, and learn about mathematics in contexts outside of mathematics. Students should be given opportunities to communicate and present mathematical ideas through the use of mathematical symbols, writing, reading, and visualization to help deepen their understanding and increase their engagement with the mathematics (274). The Representation Standard recommends that students create and use representations to organize, record, and communicate mathematical ideas; develop a repertoire of mathematical representations that can be used purposefully, flexibly, and appropriately; and use representations to model and interpret physical, social, and mathematical phenomena (280).

THE NCTM STANDARDS AND CITIZENS' ATTRIBUTES FOR THE TWENTY-FIRST CENTURY

A review of Caine and Caine's (1997) essential attributes for citizens of the twenty-first century shows they closely correspond to principles and standards for mathematics as described by the *Standards*:

♦ The "inner sense of connectedness" of the first attribute, which I translate as deep conceptual understanding, occurs from information that allows the brain to seek, extract, and make sense of patterns. It is through such a process that students, taught from a constructivist perspective, are encouraged to make sense of new information.

♦ The "strong identity" of the second attribute equates to students' confidence in working with mathematics when their ideas have been respected and valued in a classroom where equity is applied. Students' appreciation for, as wells as confidence in, mathematics is increased when the contributions of various cultures in the development of mathematics are discussed and valued.

♦ "The capacity to...deal with...uncertainty" of attribute 3 is fostered through instruction that has students solving rich problems for which there maybe no answer, one answer or multiple answers. Using technological tools and varying representations facilitate students' search for patterns to make sense of non-routine problems.

♦ The view of "how specifics relate to each other" of attribute 4 results from experiences requiring students' application of reasoning and logical thinking processes to make connections amongst ideas that may appear disparate.

♦ The "capacity to build community" is fostered, again, in classrooms where equity issues are positively addressed, cultures are valued, and students are encouraged to solve problems in cooperative groups.

In his first address as NCTM president, Lee Stiff's (2000) message, in effect, strongly supported these attributes. He said:

NCTM has always argued for a strong foundation on learning the basics. Our vision of basics, however, goes beyond mere number-crunching skills. We hope *Principles and Standards* will help educators, school boards, parents, and business leaders recognize that the new economy demands greater and more-sophisticated mathematical knowledge. 'Shopkeeper's math' alone is not enough in high tech environment. NCTM's vision of school mathematics prepares students to meet the challenges that lie ahead in a future they can't imagine. (3)

Hereafter, I refer to reformed mathematical content that is taught from the perspectives of these attributes or process standards, as the "new basics."

RESOURCES FROM NCTM

What are some strategies and resources for teachers or for preparing teachers to reach all students? NCTM's *Principles and Standards* provides recommendations. It suggests principles for the professional development of mathematics teachers (NCTM, 2000, 375–377).

WEB SITE

Online, NCTM has new support resources that include: the Illumination site (illumination.nctm.org), which provides opportunities to see Internet-based lesson plans, resources, and thoughts for reflections that are provoked by classroom video vignettes of teaching and learning; *E-Standards*, which is the electronic edition of *Principles and Standards*, allows quick and easy access to parts of the document and is available in CD-ROM or on the Web at standards.ncym. org; The MarcoPolo training program, which has professionals who will come on-site to provide training to teacher trainers.

THE PROFESSIONAL STANDARDS FOR TEACHING MATHEMATICS

Additional detailed guidelines to help teachers create a rich mathematical environment (in book form) is in the *Professional Standards for Teaching Mathematics* (NCTM 1991). It describes, to a great extent, actions needed to enable teachers to engage students in the challenging mathematics that can make students *mathematically powerful*. Students with such power can demonstrate application of the standards by their ability to explore, conjecture, reason logically, and successfully apply a number of different strategies to solve nonroutine problems.

THE ASSESSMENT STANDARDS FOR SCHOOL MATHEMATICS

Assessing this power requires multiple strategies and tools—tools that must extend beyond the traditional paper-and-pencil tests to include alternative means of assessments. The *Assessment Standards for School Mathematics* (1995), laid the ground work for *Principles and Standard's* recommendations. Evidence of student progress may come from close observation, one-on-one discussions, projects, homework, and class discourse. Such assessment is intended to help teachers make instructional decisions. *Assessment Standards* recommend that assessments should:

- Define the mathematics that all students need to know and be able to do

- Enhance mathematical learning so that students' mathematical understanding is enriched as a result of the assessment

- Promote equity so that teachers have high expectations for all students

- Be an open process so that students have a clear understanding of the assessment's goals

- Promote valid inferences about mathematics learning so that each student is given support where necessary to improve

A key change is in the use of assessment tools as a process to stimulate growth and interest in mathematics rather than as a means to separate and rank students.

THE ADDENDA BOOKS

The *Addenda* series, a group of 22 supplemental books, provide classroom-tested activities and ideas to support the *Standards* documents. However, to more closely highlight aspects of *Principles and Standards*, a new series is being published, but, as is true of the previous *Standards* documents, the old series will continue to be useful. (For a summary of K-12 school mathematics programs that model the approaches recommended by NCTM, see Appendix A.)

NATIONAL BOARD CERTIFICATION

The National Board of Professional Teaching Standards (NBPTS, (800) 228-3224) was created in 1987 to recognize teachers who are successfully implementing reform guidelines to improve student learning, establish high standards, and provide a certification process for accomplished teachers. Although the NBPTS consists of both teachers and other members of the education community, only practicing teachers may serve as assessors. To qualify for national certification, a teacher must hold a baccalaureate degree and must have held a valid teaching license for at least three years while teaching. To be nationally certified in a content area, teachers must complete a two-part performance assessment process. First, they compile a portfolio from their classroom that includes examples of their students' work, a 20-minute videotape of actual classroom teaching, other teaching artifacts and documentation, and a reflection piece that summarizes the entries and how they relate to the purposes of instruction. In mathematics, the second part is designed to allow teachers to demonstrate knowledge, skills, and abilities in four out of five domains. Currently these include algebra and functions, calculus, discrete mathematics, geometry, and statistics and data analysis. According to NBPTS (2000), teachers who have successfully completed the process, "having received the highest honor the profession has to bestow, serve as role models and spokespersons in the effort to build and to strengthen the teaching profession" (7).

2

EXEMPLARY PRACTICE: WHAT DOES IT LOOK LIKE?

While I was helping my daughter Georgine with her math homework, I asked her to explain why she chose the operation she used to solve a problem. Not only did she not know, but she also did not care. She was more interested in getting the right answer by plugging in the proper formula. This is how she was taught math, and she doesn't seem to want to change the way she learned it. I can only generalize that this is how many students are responding to attempts by teachers to create conceptual understanding. This age group is where so many students lose interest in math—just when they should be finding the beauty of it. Perhaps I shouldn't worry too much; Georgine's passion lies in social studies and literature. She is not a "math-brained" child, I guess. Are these children born, and not made that way?

Angeline,
preservice teacher

Like Georgine, many students today perceive mathematics to be a bunch of numbers that plug in to formulas to solve problems. More often than not, the problems they are asked to solve are not *their* problems, nor do they come close to something they are interested in pursuing. Georgine's experiences with mathematics are similar to those that I had as a mathematics student—the mathematics that I learned focused on finding the teacher's or the book's answer to a problem. But when I studied mathematics methods at Brooklyn College, my classmates and I explored a different kind of teaching and learning.

15

Rather than lecture us about what we needed to know, Professor Geddes invited us to experience mathematics as a dynamic discipline that sometimes required tools such as toothpicks, geoboards, or mirrors to resolve thought-provoking problems. Her definition of mathematical competence clearly went beyond numbers and computations; she included in her definition of a mathematically competent person the ability to test a hypothesis, find patterns, and communicate understanding—all of which are recommended by NCTM as essential elements of both teaching and learning mathematics.

Georgine has two problems facing too many students today: Not only does she not like mathematics, but influential teachers accept this as a natural outcome. What happens to Georgine's mathematics learning when her mom or her teacher believes that she does not have a "math brain"? The answer depends on their response. If they decide that it is acceptable for Georgine not to succeed in mathematics because she's smart in other areas and there is no reason to work to enhance her mathematical understanding, then Georgine may never change her own attitude about mathematics. On the other hand, if the belief that Georgine is not "math-brained" encourages her mother and her teacher to reason that she is strong in some areas, and to work toward connecting the mathematics that she is learning to her areas of interest in a way that makes sense to her, then Georgine has a fighting chance to understand, appreciate, and maybe even love mathematics. The reality is that every student has a unique and complex brain; our classrooms are composed of many Georgines with many varying interests and aptitudes.

The information in Chapter 1 on NCTM's recommendations for reforming curriculum, teaching, and assessment provided many ideas about what a classroom influenced by reform principles should look like to reach Georgine and other students. Not surprisingly, creating coherent lessons that promote such reform is not easy, partly because acquiring a clear vision of key elements and how they interrelate requires new ways of thinking, as well as practice, guidance, and time to evolve.

ENVISIONING REFORM-BASED CLASSROOM ENVIRONMENT

Teachers or curriculum writers must exercise caution against a limited vision of the *Principles and Standards* that might lead to a superficial or misguided application. As an example, look at the following lesson in an eighth- or ninth-grade algebra class and ask, "How are the teaching, instructional activities, and student participation different from those of a traditional classroom?"

The bell rings, and Nancy's students enter class. They quickly sit in their assigned groups of four and take out their calculators. Nancy's goal for the class is to have them model addition of integers with colored counters. She begins with a review of the properties of integers and their representations with the counters, then gives each student a package of counters and a worksheet on addition of integers. Students decide who will tackle which problem, and the groups get set to work. Nancy visits each group to monitor their progress.

This description includes many of the concepts that we associate with reform: The students are working in groups with manipulatives that include calculators, and the teacher monitors progress. How could the lesson not be reform-based? Let us take a closer look.

In her discussion of the colored counters, Nancy first defines the use of the counters: A black counter represents a positive number. Hence three black counters represent +3. Next she tells students how to add integers having the same signs and then models the example with the counters: "To add two integers with the same sign, just add the numbers and keep the sign. So, (+2) plus (+3) equals +5, and we can show this is true with the counters." She draws three black counters and adds two more blacks to show a total of five black counters, or +5. She next explains how to add when signs are different: "If the signs are different, then subtract the two numbers and take the sign of the number with the larger absolute value. So, what do we get for (–3) + (+4)?" A students gives the correct answer of +1 and Nancy then draws three white counters and four black counters on the board to verify the answer. A student asks "Why do we have to use the counters if we can get the answer by using your rules first anyway?" Nancy responds that this is just another way to do such problems. As she hands each student a sheet with exercises on addition of integers, Nancy instructs students to use the counters to show the results of their actions. Students decide who will do which problem and begin working. Some use calculators with their worksheet; when most are finished, they wait for other students to finish working. Nancy visits each group, correcting any students' errors. She assigns different students to put problems on the board when the groups finish.

Closer scrutiny shows that what looks like reformed teaching lacks key ingredients of reform. First, consider Nancy's use of manipulatives. Properly used, manipulatives provide an alternative, concrete representation that is conducive to students' initial discovery or understanding of more abstract concepts or algorithms. They are valuable when they are introduced as an integral part of a lesson to challenge students' thinking. Nancy's use of the colored counters does neither because she presents them from an algorithmic perspective. Yet colored counters are helpful to students' discovery of the rules for operations on integers. To help students discover them, Nancy would have had to connect her introduction of the colored counters to that of integers as representations of opposite situations using an appropriate model. For example, using a win-lose model translates "+3 dollars" to mean "I won three dollars," and "–2 dollars" to mean "I lost two dollars." The end results from adding is "I won three, then lost two, so I am left with only one dollar," which we represent as +1. The concept of opposite numbers follows easily since winning three dollars and then losing three dollars neutralize each other and yield zero: (+3) + (–3) = 0. Having explained this model to the class, she could then use it to develop the rules or she could introduce colored counters as a visual approach to operations with signed numbers. She could say, "Let black counters represent positive numbers, or winning, and white counters represent the negatives. How can we represent (+3)? (–2)? or 0? Let's define addition: To add two numbers is to place counters representing the numbers and then to eliminate zeros. Let's go back now and use counters to find, (+3) + (–2)." Once students eliminate zeros, they should see that only one black counter, or +1, remains as the answer (see Figure 2.1). Nancy could then have students practice adding single digit integers that they create at random. Students should be encouraged to create real-life stories to model the actions.

FIGURE 2.1 USING COLORED COUNTERS
TO MODEL ADDITION OF INTEGERS

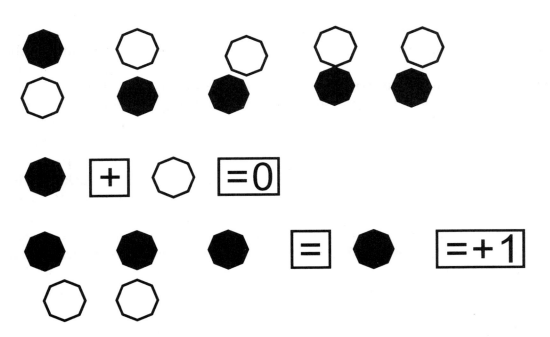

After some practice, telling students to add (+234) + (–456) without a calculator should either get a student to propose a rule, which other students should test or should it create the need for students to find a rule. From this point, students are ready to proceed in an organized manner to seek patterns. Nancy could ask students for suggestions on how to proceed or have students complete a worksheet with a sequence of problems conducive to generating the rules (Figure 2.2). Once students have discovered a rule, they should test it with several examples, and if "Chris" discovers the rule, then it becomes not the teacher's, but Chris's rule for addition of integers.

FIGURE 2.2 STRUCTURED WORKSHEET FOR DISCOVERING RULES FOR ADDITION OF INTEGERS

Directions: Work in groups. Complete the following problems with colored counters or use the win-lose story. Try to discover a pattern or rule that works for your group of problems and discuss with your group. Create three similar problems for each group using three-digit integers. Apply your rule to find the sums. See any other patterns?

GROUP 1:	GROUP 2:	GROUP 3:	GROUP 4:
(+4) + (+4) =	(–4) + (–4) =	(+4) + (–4) =	(–4) + (+4) =
(+6) + (+3) =	(–6) + (–3) =	(–6) + (+3) =	(+6) + (–3) =
(+3) + (+4) =	(–3) + (–4) =	(–3) + (+4) =	(+3) + (–4) =
(+6) + (+2) =	(–6) + (–2) =	(–6) + (+2) =	(+6) + (–2) =
(+1) + (0) =	(–1) + (0) =	(–5) + (+3) =	(+5) + (–3) =
(+2) + (+7) =	(–2) + (–7) =	(–2) + (+7) =	(+2) + (–7) =
(+4) + (+6) =	(–4) + (–6) =	(+4) + (–6) =	(–4) + (+6) =
(+6) + (+4) =	(–6) + (–4) =	(+6) + (–4) =	(–6) + (+4) =

Second, what about Nancy's use of the calculator? Students who are using it as a quick way to merely get answers to the problems are using it inappropriately. However, those using it to check their guesses for addition of large integers or to find new patterns are engaging the full power of the calculator to promote higher thinking. Nancy's arrangement of students' seats in pairs versus in rows of desks may be conducive to small group processing of ideas. Note that the *Principles and Standards* recommends that students work in small groups because "This approach is often very effective with students in the middle grades because they can try out their ideas in the relative privacy of a small group before opening themselves up to the entire class" (NCTM 2000, 272). In Nancy's groups, students worked individually applying her rules, therefore, there was little motivation for group members to share ideas, even though the small group size would have made that easy to achieve. Furthermore, it was Nancy, not group members, who judged the correctness of answers and determined who would report answers on the board: She did not try to assess students' understanding, pose questions to provoke further thinking, or suggest to students that they enlist the help of other group members.

Wheately (1991) succinctly describes key elements for facilitating small groups. He writes that in preparing for classroom instruction,

> …a teacher selects tasks which have a high probability of being problematical for students—tasks which may cause students to find a problem. Secondly, the students work on these tasks in small groups. During this time the teacher attempts to convey collaborative work as a goal. Finally, the class is convened as a whole for a time of sharing. Groups present their solutions to the class, not to the teacher, for discussion. The role of the teacher in these discussions is that of facilitator, and every effort is made to be nonjudgmental and encouraging" (15–16).

Was Nancy's approach bad? No, there might have been some educational gains for some students. Learners construct their own knowledge at all times and in all types of situations, but different instructional approaches may influence the quality and content accuracy of the construction. The fact that students faced each other in small groups rather than in rows looking at each others' backs surely promoted some worthwhile discussion among students, but the amount and quality of their interchange, in terms of learning the mathematics involved, would undoubtedly have been increased had Nancy designed the assignment to challenge students. Although the colored counters were not applied in the best way to enhance the students' ability to make connections between multiple representations, they still provided an alternative view of operations with integer terms, and they may have helped some students better understand the mathematics. Nancy also had students present their answers, thus opening an opportunity for students to share their thinking and summarize ideas.

I surmise that Nancy's perception of teaching mathematics is one that relies on teacher control or is conceptually rule driven. She probably has had little experience using various tools, such as manipulatives, to guide exploratory activities. However, the fact that she has elements that are conducive to reform activities in her class indicates that she is trying to embrace different approaches to teaching. Her instruction and choice of activities are those of a teacher in transition to a reform-based teaching approach. A clearer vision of what the *Principles and Standards* recommend is a key to her success at moving forward with the transition.

Now let's consider a typical classroom of 30 students who are sitting in straight rows and busily working individually on a worksheet. Claudette, the teacher, stands at the front of the room or occasionally circulates about and looks over their shoulders. Is she teaching from a reform perspective? Maybe. It depends on what the worksheet requires and whether students have opportunities to learn in other ways described by the *Principles and Standards* on other days. Suppose Claudette's goal is for students to apply the heuristic "think of a simpler problem" to problems that are not routine. Below are the examples on the worksheet:

1. Find the last two digits of $11^{20} - 1$

2. Determine a rule for finding the following sum:

   $1^2 - 2^2 + 3^2 - 4^2 + 5^2 \ldots + 1999^2$

3. Be prepared to explain to the class your strategies for getting your answers.

The sheet is not the "drill-and-kill" variety. It requires students to apply sound problem-solving heuristics to problems that are suitable for individual work. Further, the third question will promote the sharing of students' ideas and discourse. If she occasionally varies her teaching style, she may be teaching from a reform-based perspective.

The two previous examples show that labeling an activity or class as reform-based or not requires close scrutiny of the work students do, how they do it, and whether a single teaching method is used exclusively. Let's consider the revisions I made to Nancy's lesson and ask one final question: Is it now reformed? Some would say yes, somewhat, but ask, "What about a real-life application out of which the need to compute signed numbers arises? Why not have students do individual explorations first before going into groups?" My point is that probably many of us are traditional teachers in transition and are also at various levels of understanding what the *Principles and Standards* imply. Further, we all come to the table with different experiences, expertise, and expectations. Instrumental to helping us move closer to reformed-based teaching are "opportunities to reflect on and refine instructional practice—during class and outside of class, alone and with others..." (NCTM 2000, 19).

EXEMPLARY PRACTICES

There are exemplary practices that clearly demonstrate best practices for teaching and learning for understanding. For example, the teachers profiled in this book:

- ◆ Engage students in challenging, mathematically appropriate tasks that make sense to students.

- ◆ Create a classroom atmosphere conducive to discourse that encourages students' alternative conjectures, approaches, and explanations.

- ◆ Use appropriate tools, cooperative group work, and individual instruction to accommodate students with different learning styles.

- ◆ Use alternative assessment methods to assess students and guide their instruction.

- ◆ Collaborate with colleagues and pursue other professional development activities to support or improve their practice.

Do any of the teachers lecture at times? Sure. Many of us learned from lectures (of course, how well we understood what we learned is subject to debate). Past NCTM President Gail Burrill (Burrill 1998) elaborates on perceptions to avoid when teachers attempt to implement reform:

We must avoid misinterpretations such as: everything must be done in cooperative groups; decreased emphasis means none at all; every answer to every problem has to be explained in writing; the teacher is only a guide; every problem has to involve the real world; computational algorithms are not allowed; students should never practice; and manipulatives are the basis for all learning. The challenge is to make choices about content and teaching based on what we can do to enable students to learn.

In his article in the March 1997 issue of *The Mathematics Teacher* about applying common sense when one is implementing the *Principles and Standards*, classroom teacher Mark Saul echoes similar views:

So what constitutes "real world?" As a classroom teacher, I have an operational definition: If it holds my students' attention, it is in their real world. If it does not, it is not. My job is to bring more mathematical thinking into their "real world."...What about technology? Should we not use calculators "at all times"? Well, no. We should be free to choose when to use them and when not to. (Saul 1997, 183)

As mathematics educators, we know very well to be wary of universal statements, such as "for all x, y is true." Both Burrill and Saul are recommending that we be mindful that its negation, "there is an x for which y is false," is often true when x represents students in our class and y represents a statement about the effectiveness of a specific activity or method. In essence, they are suggesting that we think as self-directed learners in the activities and strategies we select to reach our students. If we do not heed their caution, I fear that education will continue to be entangled in radical movements that stress one philosophical stance over another. Keeping our focus on all students' learning illuminates the fact that our students are too diverse to be neatly served by instructional methods labeled, "Use me all the time!" The key reflective question that should guide whatever approach we take is, how can we best facilitate students' understanding of the mathematical content in a meaningful way that contributes to their success in the twenty-first century? Our best practices should align with the answers to that question.

3

SUSAN MERCER AND TAMMRA DETVILER: PATTERNS FOR CALCULATING PERIMETER AND VOLUME OF OPEN PENTOMINOES

I currently teach at a multilingual, minority, low socioeconomic school where students are heterogeneously grouped by grade level. I have done this project with my three seventh-grade classes, two of which are inclusion classes in which I team-teach with the Special Education teacher. I used this unit with Mainstream, Limited English Proficiency, and Special Education students. While the mathematical skills of my students range from very good to very poor, this unit allowed all students to feel successful because it provided them with an opportunity to experience important mathematics concepts through participation in a challenging project.

Susan Mercer and
Tammra Detviler,
Spurgeon Intermediate
School, Santa Ana, CA

In Susan Mercer's four seventh-grade classes, students have a wide range of math skills and English proficiency. In addition to mainstream students, one of her classes has special education students from a Special Day Class (SDC), and special education students needing support from a Resource Specialist Program (RSP) class. The remainder of her classes consists of students classified as having Limited English Proficient (LEP). Susan describes her special students as follows:

> The SDC students receive special education services and support during 51 percent or more of the school day. The SDC students' disabilities range from Specific Learning Disabled to Mild Mental Retardation, Seriously Emotionally Disturbed, and Traumatic Brain Injured. RSP students receive special education services and support during less than 50 percent of the school day. Their disability is diagnosed as Specific Learning Disabled.

Susan's inclusion class has 42 students. When I first read her profile, I immediately assumed she was a special education teacher. How else could she teach such a large number of special students without training? My communications with her quickly revealed that I had overlooked the process of team teaching in inclusion classes.

How did she get into this situation? Susan says that it started when the special education teacher, Tammra Detviler, approached her about the possibility of including special students in one of the mainstream math classes. Tammra felt math was not her strongest subject, and she wanted her students to experience a rich, challenging, hands-on mathematics program. In addition, she wanted someone who could teach the mathematics content in a clear and exciting manner. Because they worked together when Susan had a special student the year before, Tammra understood and embraced Susan's teaching style and philosophy.

Challenging such a mix of kids has to be an incredible undertaking. Susan admits to this and writes, "Preparing, planning, and presenting a project that will challenge my high math skill students in the class, while providing the tools for the low math skill students to work on the project is not an easy task." Susan and Tammra's agreement to merge their classes opened doors to new experiences in teaching and learning for both of them.

PARTITIONING THE WORK

How to manage and divide the work so that both teachers would work efficiently while providing a cohesive program for the students had to be well thought out by the teachers. During class, they decided that Susan would present the day's lesson no differently than she does in her mainstream classes. Tammra would take attendance, sign readmit slips, and hand out material to students who have been absent. After a lesson, she would also check for understanding by asking students what they had to do, what were the next steps, and what needed to be accomplished during the period. Once students started working on their projects, both Susan and Tammra would move around the classroom helping students, answering questions, and assessing students' progress in the unit. The arrangement of desks and selection of student groups were also important. Students with varying abilities in math and English skills would be seated around hexagonal tables in groups of four or five general education students and one or two special education students. Through trial and error, the teachers learned that the best ratio was one special education student to two or three general education students. Tammra's expertise was especially valuable for determining who sat with whom.

Because of the range of student's skills, Susan is careful to present the math content in ways conducive to giving every student a chance to access the core math curriculum. At her school, this curriculum is a set of math concepts, projects, and units that the math department as a whole has decided to teach at each grade level.

ENGAGING THE STUDENTS

Because she believes that projects have to be interesting, age-appropriate, and challenging, Susan uses a number of group activities to create a project. She feels it is also important for projects to contain some level of confusion and a small dose of frustration to increase the chances that students complete the project with a sense of pride in their efforts and final product. She applied these principles to create a project using pentominoes to engage students in the process of making connections among the concepts of perimeter, area, volume, and graphs of these measurements. "I developed the pentominoes project so that all of my students could experience some level of success," she says. Before beginning the projects, students completed a short, hands-on unit that introduced the concepts of area and perimeter.

Susan starts the project by asking students for a definition of pentominoes. Generally no one knows, so she explains that a pentomino is a shape formed by putting five squares together so that the sides touch but do not overlap. She shows Figure 3.1 on the overhead for students to examine and gives each table a bin with 100 one-inch tiles.

Working in pairs or groups of three, students use the tiles and try to find as many pentominoes as possible. Once students have completed this first task, Susan asks them to share their results. Her next assignment requires that pentominoes be replicated on graph paper, cut, and then folded to find those that make an open box. She observes that, while some students try to fold all of them, others see conditions necessary for this to happen so that they judge some drawings as unlikely to form a box. Students paste those pentominoes that fold into an open box on one side of a paper, and those that don't, on the other side.

FIGURE 3.1 TWO PENTOMINOES OF LENGTH 1

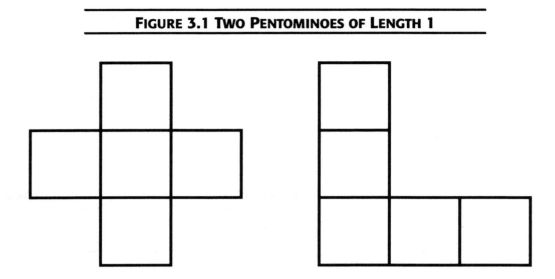

Because it is easy and interesting to fold and test the pentominoes, this is a fun exercise for students. Susan asks one student from each group to share discoveries with the rest of the class. The activity creates a lively discussion among students because some of the suggestions for patterns are viewed with skepticism by others, who then challenge and test the patterns. The successful student will have eight pentominoes folding into an open box and four that do not. Susan next shows Figure 3.2 on the overhead and asks the class to compare and contrast it to Figure 3.1.

FIGURE 3.2 A PENTOMINO OF LENGTH 2

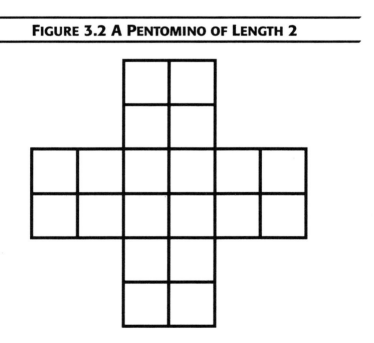

She welcomes students' comment on the similarities and differences and, if necessary, directs their attention to length of sides of the largest square in each. She explains that because each one of the five squares that make the pentominoes of Figure 3.1 has a length of 1 inch, they are called pentominoes of length 1. "What do you think a pentomino of length 2 would look like?" Students have no trouble responding that a pentomino of length 2 would have each one of the five squares having a length of 2 inches, as shown in Figure 3.2.

In the next part of the project, Susan asks the class to select one of the pentominoes that folds into an open box, replicate it once more on graph paper and cut it out. Now they have to use 1-inch tiles to construct a pentomino with a length of 2 that is similar to the one they have selected. This is not an easy task because many students double one dimension of the pentomino but omit to do so for all sides. Susan makes sure that all students successfully complete this task before continuing. Once successful, she instructs students to replicate the pentomino of length 2 on graph paper, cut it out, and compute its area and perimeter.

Her next goal is to introduce the concept of volume. She begins with questions to review and extend their measurement concepts: "How did we find the area and perimeter of Figure 3.2? How could we measure how much space it encloses or how much milk a container of its size might hold? How many dimensions does a square have? If we use a square to measure a flat surface, what shape do you think we might use to measure three dimensions? What name is given to this type of measurement?" She distributes 1-inch cubes to students and asks them to find the volume of the open box formed by the pentomino of length 1. Once all the students have checked that the volume is 1 cubic inch, Susan asks the students to estimate the volume of the open box formed with a pentomino of length 2. She gives each group 10 cubes to verify answers. Students compare answers, and once Susan assesses that students have the idea, she instructs the class to prepare a poster to display pentominoes of length 1, 2, 3, 4, and 5, in order. The poster must include the length, perimeter, area, and volume of each pentomino, a written explanation on how calculation was performed, and the following graphs with a summary of what the graphs show: length-perimeter, length-area, and length-volume. Because the task is complex, students assign parts to each group member and begin work. As Susan moves about the class, she observes groups taking different approaches to tackling the various parts. Some groups calculate the perimeter of the five pentominoes, then the area, and finally, the volume. Other groups start by calculating the perimeter, area, and volume of the pentomino of length 2, then the pentomino of length 3, and so on. Some students make the graphs, while others explain in writing how to calculate each part. Students work in pairs and then share, compare, and discuss results. She also observes the level of students' engagement because she wants to assure that students at all levels have input into the final product. As she monitors the participation of her mainstream, LEP, and SDC students, she notices the roles they play. She writes, "One of my SDC students could not explain in writing, nor could she come up with procedures. But she decided she could and would number all the 1-inch squares to show the area of each pentomino. Her contribution was valued by members of the group and she felt proud of her participation." This example illustrates how use of manipulatives can play a significant role in students' participation: While the high math skill students look at the result of the manipulative work to suggest patterns and procedures for describing the process, others, not as quick, use the manipulatives to continue to solve the problems at their level of understanding.

Susan's goal, however, is to provide opportunities for *all* of her students to move toward a higher abstract level of thinking. To support this she includes at least one problem that is not solvable with manipulatives. For example, while the volume for the pentomino of length 5 requires 125 cubes, each group received only 100. Susan says,

> Students thus have to problem solve and calculate the volume based on what they have already learned. I do not help them because I want them to construct their own procedures to calculate the volume by sharing ideas and possibilities with their group. During the project it is very interesting to observe how students moved from using manipulatives to the abstract so as to calculate the volume. With the pentominoes of length 2 and 3, students used the 1-inch cubes to calculate the volume because it was easy. However, as the task of "filling" the pentominoes with cubes got harder and more complicated, students started looking for shortcuts or procedures that would allow them to find an answer without using the cubes. As they found procedures, students would test their "theories" using the cubes, but I also observed students getting excited for having "discovered" procedures they were able to explain. At the end of the project, most groups were finding the volume by calculating the base then multiplying by number of "layers going up." In one of the groups having one LEP Vietnamese student and five LEP Hispanic students, the Vietnamese student "informed" the rest that he knew how to calculate the volume without cubes, but the Hispanic students were not convinced so they checked the answers with the cubes. After doing four of the five pentominoes, they were convinced and started using the algorithm.

As students brainstorm in their groups, Susan and Tammra walk around asking questions to assess understanding of the concepts. In some cases Susan asks specific questions, while in others she just listens to discussions. This is a critical aspect of this project because it enables her to challenge every student to perform at his/her highest potential. For example, one group calls her to say that they find both area and volume of the pentomino of length 5 to have the same number, 125. Susan takes a cursory glance at their numbers and replies, "I don't believe that is possible," and she leaves. The students look puzzled and begin talking among themselves. After a while they call her back and say that it *is* possible. They explain that they have 5 big squares, each having an area of 25 sq. in. These squares therefore have a total area of 125 sq. in. To find the volume, take the 25 cubes from the base, multiply it by 5 "layers going up," and that yields 125. In both cases, they conclude you have five squares with 25 squares or cubes each. Susan, writes, "They thus presented a valid, convincing argument for their theory and felt proud about it. Questioning what students are doing not only forces them to think and make connections, but it also forces them to be able to explain their thinking not only to me but to each other."

On the day before groups finish posters, Susan reminds them to present their work in an organized, clear, and creative manner, using correct mathematical symbols and vocabulary. The next day, as groups present their posters, Susan interrupts the presentations periodically to check and verify that the class understands the reasoning supporting the calculations. The presentations lead to discussions, which Susan considers the most interesting and productive part of the project since every group has the opportunity to listen and see different strategies for calculating the solving problem. For example, one group calculates the area by counting the number of 1-inch squares in the pentomino and then multiplying by 5; to that, another group adds that they do the same but calculate the area of one square by multiplying length times width; to this a different group adds they just multiply the length times itself because the length and width are the same. As they arise, Susan discusses these strategies with the class so that they may be better able to describe and generalize patterns. She carefully notes them on the board and reminds the class likewise to keep good notes.

Once every group has shared, Susan asks the class to consider solving two final problems: "Calculate, *without cutting out*, the perimeter, area, and volume of a pentomino with a length of 6 and 7 inches." To help students, she leads the class in a summary discussion of the strategies they used to solve the other problems, and she asks students to seek patterns that might be useful. These problems are yet another attempt on her part to get more students to move toward abstract levels as she assesses their progress.

DISCUSSION BETWEEN COLLEAGUES

What made you decide on this project?

> During 1995/1996, I took a math course through my school district provided by California State University, Fullerton. The objective of the class was to improve the math content knowledge of teachers, and

it also provided strategies and methodologies based on the NCTM *Standards*. During the course I asked the instructor if she knew of any projects to teach perimeter, area, and volume, and she suggested this project. I tried it with my students, and they really enjoyed and learned from it.

How often do you use projects with your students?

I do projects all year with my seventh graders—two or three projects a trimester. I use the projects to uncover most of the content seventh graders are expected to learn. I organize my seventh-grade classes in the following way: On Monday I assign homework that consists of two or three problems for students to practice and/or review what we have done in class and/or practice basic skills; Tuesdays, Wednesdays, and Thursdays, students work on projects; Fridays we review, share, and grade the homework assignment. Our periods are 45 minutes long, except for Fridays, with 30-minute periods.

Would you make any changes to the project?

I think this was a great project, and I plan to do it again this year. In general I think the project went really well, although I would add two modifications: (1) I would ask students to generalize the perimeter, area, and volume of a pentomino of length L because I think many could have done this; and (2) in addition to the tables and graphs I would ask students to spend 20 minutes of quiet time to write and/or think about summarizing what they did and learned during the pentomino project. Questions I would ask include: (1) In your own words, explain how to calculate the perimeter, area, and volume of a pentomino; (2) Explain the difference between area and volume, or between perimeter and area; (3) What is the difference between a square and a cube? (4) Which one is used to measure area? volume? Why? (5) What did you learn during this unit? For those SDC students struggling with writing, Tammra and I can help them write their thoughts. Students need to individually reflect on what they have done and learned in order to assess their own individual growth.

Do you have extensions for the project?

One of the fun parts occurs when asking the groups questions about their graphs, such as, "What happens if a pentomino has a length of 0? How might you show a pentomino of length 0 in the graphs? How can you use the graph to find the perimeter, area, or volume of a pentomino of length 1.5 inches? How might we generalize to a pentomino of length of n?" Students were very intrigued with the graphs and all the information they provided.

How do you assess the project?

Because of the wide range of math skills in my classes, I assess the students based on what they have learned rather than focusing on what they still have to learn. Every student has the opportunity to show in writing and orally what he or she knows. Special education students are assessed based on their Individual Educational Plan (IEP). The level and depth of their understanding depends on their learning disability. Students are valued based on their individual growth, rather than compared to each other, but students are expected to perform to their highest potential. Those students who can't write what they have learned, can tell us or show us using manipulatives. Finishing a project is the responsibility of every member of the group, and students know that they are assessed on the quality and depth of their work. Since assessment is based on individual growth and differences are respected, my classroom environment is one of collaboration rather than competition—students helping students and groups helping each other.

During the project Tammra and I walk around asking questions to assess the understanding of the different concepts. In some cases, we ask specific questions, in others we just listen to the discussion among the students in a group.

Each student receives two grades for this project. One grade is a group grade based on completeness, creativity, and depth illustrated in the posters. The second grade is an individual grade based on individual class participation during the project and homework. For homework, each student prepares a report that includes the pentominoes of length 1 and 2, and the table with the length, perimeter, area, and volume for the different pentominoes and the graphs.

Have you always taught from a reformed perspective?

Yes. I went to a university where classes were taught in a progressive manner and all the lessons and units I prepared and presented were based on the California Mathematics Frameworks and NCTM *Standards*. This allowed me to understand and read the *California Math Framework* before I started teaching. In addition, I started working as a math teacher in the Quasar project, a Ford Foundation grant managed by the Learning Research and Development Center housed at the University of Pittsburgh. This was a great advantage as I was involved in developing the scope and sequence of the site math curriculum as well as taking advantage of the staff development opportunities provided by the grant. Furthermore, I was able to work with members of the mathematics department in looking for ways to improve the instruction, delivery, and assessment of the math content for all students.

Does your school have a different curriculum for special education students? If not, what guidance do special education teachers get?

At Spurgeon, we try to align our curriculum to the NCTM Standards. We try to provide special education students with the same core curriculum that we use with limited proficiency and general education students. To do so, the special education teachers and instructional assistants are included in the math department and participate in all staff development opportunities. On curriculum planning days, input from special education teachers is valued because they suggest modifications and varying strategies for presenting the curriculum. The suggestions are beneficial to all students. Thus, we do not "dumb down" the curriculum for the special education students but modify it in a way to make it accessible for ALL students.

How does your inclusion model work?

Tammra has 16 SDC students on her caseload divided between two math classes. In our inclusion class we have 10 SDC students and eight RSP students from a different special education class. The other six of Tammra's students attend a sixth-grade math class, occurring during the same period as my class and taught by an RSP teacher. In summary, there are two math inclusion classes taking place during the same period where a math teacher team-teaches with a special education teacher.

The benefits of placing special education students in general education classes are numerous: increased access to core curriculum, increased exposure to appropriate peer models, higher expectations of special education students by teachers and by themselves, increased self-esteem, increased motivation and assumption of responsibility, and increased independence. The general education students benefit in the following way: increased appreciation of individual differences; exposure to an additional teacher and teaching style; increased opportunity to participate in a variety of learning experiences such as peer tutoring, cooperative learning, presenting material to other students; and lower student/teacher ratio.

Describe further how you and Tammra team-teach.

From the first day of class, Tammra and I introduce ourselves as the teachers. We both teach, we are both in the class, and we both help all students. From the beginning it is our rules, our expectations, and our class, and these concepts are conveyed very clearly to all students. When we plan the curriculum, we use each other's strengths. Since I am the mathematics specialist, I develop the units and projects that we will use with all my seventh-grade classes. I decide what is going to be taught, which manipulatives are to be used, what the end product should be, and what homework is to be assigned. Tammra, as the special education specialist, looks at the units to develop modifications that will allow all students to be successful. First, she makes sure she clearly understands the objectives of the unit and the content to be taught. Then, if necessary, she suggests modifications. For us, modifications means changing the sequence and/or timing, modifying the way a topic is introduced, suggesting a different way to explain the same concept, reducing the amount of work done by students without missing the content, developing extensions, making posters that will help students clarify math terms or concepts, and so forth. I implement these modifications not only in my inclusion class but also with all students in all my classes.

What accommodations did you have to make to team teach with Tamara?

At the beginning it was strange and slightly uncomfortable to have two teachers in the classroom. The first couple of weeks Tammra and I were very aware of each other and what the other one did. After each period we would meet and talk about the class and the students. We shared and commented about the lesson, what worked, concerns, and the next days' lessons. This really helped us to get to know each other and to feel comfortable with the teaming situation.

What benefits have you gained from team teaching with Tammra?

Overall, team teaching is a great experience. Tammra and I benefited in several ways. It provided the opportunity to collaborate with another professional; it provided the opportunity to share problems and successes; it reduced the sense of isolation; it increased access to strategies that work with at risk students and easily extendable to all students; and it increased our understanding of the needs and abilities of special education students.

One of the compliments Tammra and I often get from visitors is the fact that they cannot identify the special education students by just observing students working.

What tips can you give readers about working in inclusive classes?

First, find someone you can work with within the Special Education department. Someone who has the same philosophy as you do. In my case, both Tammra and I believe: "ALL children can learn." Ideally, this person should have a different set of strengths than you do, in order to complement each other's teaching strategies, techniques, and content knowledge. Second, it is important to understand that team teaching and inclusion is a process—it does not occur overnight. It takes time and a lot of communication. Sharing and being honest with what worked and what did not work, and being reflective as a team on what can be done to improve, are essential. For example, at the beginning Tammra was very uncomfortable getting things from my cabinet because she felt she was intruding. Because she was honest and shared her feelings with me, we came up with suggestions to make her comfortable.

Tamara, what benefits have you gained from teaching with Susan?

A great benefit is really directed toward my students. While mathematics is my weakest area and therefore posed a worrisome task for me to create good lessons, Susan is an expert at teaching mathematics in an understandable manner. She knows how to make it a fun and learning experience for all students. I have also benefited by observing how she creates activities to engage students in higher-order thinking. She and I determined who would do what at the beginning of a class, and we have learned to work with each other on a daily basis to share successes, refine lessons, and to simply enjoy conversations and humor with each other. The most appreciated aspect of all, however, is that Susan welcomed not only me, but also all my of students to share her class.

COMMENTARY

In agreement with NCTM's view on students' potential to succeed in doing mathematics, Shrine and others from the National Center on Educational Outcomes, write: "All students have the right, and must receive the opportunity, to learn to meet high, rigorous content *Standards*. 'ALL' can mean 'all'" (Shriner, Ysseldyke, and Honetschalger 1994, 41). Specifically, which students comprise the "All?" Stiff (1993) writes,

> NCTM's position is clear. By every child is meant—
>
> 1. Students who have been denied access in any way to educational opportunities as those who have not:
>
> 2. Students who are African American, Hispanic, American Indian, and other minorities as well as those who are considered to be part of the majority;
>
> 3. Students who are female as well those who are male;
>
> 4. Students who have not been successful as well as those who have been successful in school and in mathematics (3).

Susan and Tammra's efforts show how to put this vision into practice. Merging classes not only improved their teaching and student learning, but also sent a positive statement to students that in effect said: "You can all do the good, worthwhile, and enjoyable mathematics described by NCTM." Students were given a challenging, interesting, and complex problem to solve with no quick algorithmic solution (mathematics as problem solving). They were allowed to use manipulatives to look for patterns in order to simplify problems and to lead to generalized procedures for calculating problems without manipulatives (mathematical tools). Thus, even though some students could have counted squares and cubes to calculate the area and volume, some looked for patterns and procedures to find shortcuts for calculating the area and volume of the pentominoes (mathematics as reasoning). Students recorded the perimeter, area, and volume of the different pentominoes on a table, looked for patterns, and graphed the information (patterns, functions, algebra, representation). They had the opportunity to explore two- and three-dimensional shapes while they informally explored the growth of perimeter, area, and volume and by so doing were able to differentiate among these concepts (geometry, spatial visualization, measurement). Students had to work in teams, discuss, communicate ideas and procedures to solve the problem, and present their work in an organized, clear, and creative manner (mathematics as communication).

With respect to the students with LD in the class, the teachers' approach positively addresses a concern Usiskin (1993) presents in response to the number of students labeled as LD in the United States. He writes:

> There are students who are truly learning disabled, but it also happens that we in the United States seem to have the greatest percentages of learning disabled students in the world…can you imagine going through school with an asterisk by your name that announces to

the world of teachers that you have a learning problem? How many students are learning disabled because they are doing what is expected of them (14)?

Tammra and Susan's class kept opportunities open for all students to show their strength. Every student was smart enough to be given the same problem, and everyone was expected to succeed. Getting students to believe this fact required a unit that encouraged a variety of problem-solving approaches ranging from concrete experiences to abstract ones, creation of heterogeneous groups, and the teachers' circulation among students to assure students' understanding and participation in the major components of the lesson. The heterogeneous groups played a key role. Davidson and Hammerman's (1993) work describes the successful dynamics at play in such groups:

> When we work with children in heterogeneous groups, our discussions tend to focus somewhere above the middle and upper range of students. Because our content is "rich" there is appropriate work for "lower" students to do as well. But it also means that the "lower" students are exposed to more sophisticated ideas than they would in a homogenous classroom...they become intrigued by ideas, even friendly with them, and so are motivated to explore even difficult concepts in a way appropriate for them (204).

The instructional approach used by Tammra and Susan is also recommended by Cuevas (1991) for teaching students with Limited English Proficiency. He suggests that the following instructional strategies be followed: (1) give students opportunities to clarify key terms and words; (2) create a classroom environment conducive their participation; (3) provide opportunities for students to examine and discuss problem-solving processes; (4) create opportunities for students to reflect on the main points of a lesson; (5) offer opportunities for students to talk about mathematics with one another, using their native language as well as English (186–188). All of his suggestions were applied successfully in the inclusion class. Because they are also consistent with what NCTM recommends for all students, it is not surprising that Susan uses these strategies with her mainstream classes. This begs the question: Is it enough to just do good teaching in classes of students having special needs? Khisty (1997) writes:

> We cannot assume that good teaching is simply good teaching and then carry on with the same set of assumptions about language, ethnicity, and class....Much of improved learning of Latinos and other ethnic- and language-minority students rests with teachers dispelling the myths of "disadvantages" among students, understanding how students' characteristics can be learning capital, and using abundant

resources and strategies to accommodate students' unique needs instead of excluding them. It also rests with teachers' understanding the need to work collaboratively with specialists in bilingual and ESL programs (100).

Susan and Tammra's collaboration is thus at the heart of their success in teaching special needs students. They employed what may be their most valuable tool: each other. Not only did students receive the best instruction resulting from this collaboration, but they saw their teachers similarly collaborating to promote an interesting and worthwhile endeavor: improving student learning. According to Goleman (1995), the teachers and students experienced that

> [w]henever people come together to collaborate, whether it be in an executive planning meeting or as a team working toward a shared product, there is a very real sense in which they have a group IQ, the sum total of the talents and skills of all those involved. And how well they accomplish their task will be determined by how high that IQ is. The single most important element in group intelligence, it turns out, is not the average IQ in the academic sense, but rather in terms of emotional intelligence. The key to high group IQ is social harmony. It is the ability to harmonize that, all other things being equal, will make one group especially talented, productive, and successful and another with members whose talent and skill are equal in other regards, do poorly (160).

As Susan and Tammra continue to increase their group IQ, they create activities to similarly increase their students' group IQ. What a wonderful learning spiral!

CONTACT

Susan Mercer and Tammra Detviler
Spurgeon Intermediate School
2701 W. Fifth St.
Santa Ana, CA 92703
Phone: (714) 480-2200
E-mail: mercersue@aol.com
            tammradtld@aol.com

UNIT OVERVIEW: PATTERNS FOR CALCULATING PERIMETER AND VOLUME OF OPEN BOXES

Aim: How can we find patterns to help calculate the area, perimeter, and volumes of open boxes?

Objectives: Students will informally develop algorithms and procedures to calculate perimeter, area, and volume of open boxes by finding and using patterns developed from pentominoes of two and three dimensions.

Grade Levels: 7–8

Source: Original

Number of 50-minute periods: 10 (1 day for introduction; 1 for perimeter of length 2; 3 to prepare posters; 2 for presentation; 2 for extensions, reflections, and reports)

Mathematics Principles and Standards Assessed:

- Principles for equity, curriculum, teaching, learning, assessment, technology

- Mathematics as problem solving, communication, reasoning, representation

- Number and operations

- Algebra

- Measurement

- Geometry

Mathematical Concepts: Students will compute perimeter, area, and volume for open boxes.

Materials and Tools:

- Bin with 100 tiles per group

- Bin with 100 one-inch cubes

- One scotch tape per group

- Graph paper, markers, and rulers

- Poster board

Management Procedures:

- Time Line

  - 1 day for introduction

- 1 day for perimeter of length 2
- 3 days to prepare posters
- 2 days for presentation
- 2 days for extensions reflections and reports

♦ **Procedures:**

- Part 1. Ask students to find different pentominoes using 1-inch tiles, then draw on and cut out from graph paper. Have them paste on one side of a page all the pentominoes that fold into an open box and on the other side of the page the ones that do not.

- Part 2. Have students cut out the pentominoes of length 3, 4, and 5 from graph paper.

- Part 3. For each pentomino of length 1, 2, 3, 4, and 5, have students find the perimeter, area, and volume and complete a summary table. Discuss how to find volume.

- Part 4. Have students make a poster that shows the length, perimeter, area, and volume for the selected pentominoes. Include following graphs: length-perimeter, length-area, and length-volume, with a description of each graph.

- Part 5. Have students calculate, without cutting out, the perimeter, area, and volume of a pentomino with a length of 6 and 7 inches.

Assessment: Assign a group grade based on the completeness, creativity, and depth of the poster. Assess individual students based on individual class participation during the project and homework.

4

LAURA MULLEN: LEARNING STYLES AND PROBLEM SOLVING

Oh, it would be so nice if while doing all of this teaching I had more time to write about it in a form meaningful to others. I believe the NCTM Standards are here to provoke mathematical thinking in students and teachers alike. I chose this lesson because it generates a lot of information about the students' mathematical thinking that both students and teachers can discuss. In addition, this mathematical problem allows for many solutions that can be represented in various forms so that I can help all my students reach a level of success while engaging in lots of decision making.

Lauren Mullen,
7/8th-grade teacher,
Mount Nittany Middle
School, Pennsylvania

Imagine giving students a take home test on the very first day of class! How can any teacher do this and not promote or enhance whatever math anxiety students already have? Surprisingly, Laura's test is one that students appreciate and consider valuable to them during and years after her mathematics class. It is not a test to determine whether a student has a good or poor understanding of mathematics concepts. Rather, it consists of only one problem that aims to gain information on students' team role preferences as well students' approaches to solving a nonroutine problem. The staircase problem (Figure 4.2, p. 47), typically found in lessons on problem solving, focuses on getting students to generate a table or list of examples for the purpose of finding a pattern or formula. Laura's application of this problem, however, has an important second focus. She writes,

I use this problem on the first day of school to provide me with some personal information about my students' learning styles and cooperation gifts. Students often incorrectly assume that working as a group and a team are the same thing. However, when students are doing group work I see them working toward a group goal. They focus on solving a problem where they share ideas on interpretation, strategy, and answers, while they do individual and parallel work. On the other hand, students working as a team assume responsibility for solving a problem by sharing tasks and contributing a unique aspect to the process and presentation of the solution. I associate teamwork with role-playing and it is that which I hope to develop in my students so that they grow in their approaches to solving problems.

PREPARATORY ACTIVITIES

To help students differentiate between groups and teams so that they can better play their team roles, Laura begins by explaining that not everyone is the quarterback on a football team, but everyone reads the same books in a book-sharing group. Also, she says that a team cannot go on without all its players, but a group can. She asks students to think of some elements that are critical to helping a team function successfully. Students mention that while members need to be patient with and respectful of each other, they must also be willing to contribute. "Contribution is certainly a critical factor. Knowing or understanding how you prefer to tackle a job is important to getting your best level of participation," she adds. "What are the ways that best help you to learn? How many of you learn best by reading the information from a book? How many of you prefer to try things out for yourselves?" she asks. Wanting to help students learn more about their preferred styles of learning so they can develop skills necessary to successful teamwork, she engages students in a discussion on the seven styles of learning. Some good resources that she uses are the books, *Multiple Intelligences in the Classroom*, by Thomas Armstrong (Association for Supervision and Curriculum Development, phone: (800) 933-2723), and *Strengthening English and Social Studies Instruction*, by Roger Taylor (Bureau of Education & Research, Bellevue WA, phone: (800) 735-3503).

Before assigning the staircase problem, she briefly describes each intelligence and asks students to identify which of seven multiple intelligences (MI) they think best promotes their own learning. She distributes a handout adapted from Taylor's book that describes each type:

1. Linguistic Learners like to read, write, and tell stories. They are good at memorizing details and learn best by saying, hearing, and seeing words. Taylor calls them "Word Players."

2. Logical/Mathematical Learners like to experiment and explore patterns. They are good at mathematics and problem solving, and learn best by categorizing and working with abstract relationships. Taylor calls them "Questioners."

3. Spatial Learners like to build things and look at pictures or movies. They are good at imagining things and making sense of charts and puzzles. They learn best by visualizing and working with pictures. Taylor calls them "Visualizers."

4. Musical Learners like singing or playing a musical instrument. They are good at keeping time and picking up sounds. They learn best through activities involving aspects of music such as melody and rhythm. Taylor calls them "Music Lovers."

5. Bodily/Kinesthetic Learners like talking or moving around. They are good at physical activities like sports, dance, and acting. They learn best by touching, moving, and interacting with space. Taylor calls them "Movers."

6. Interpersonal Learners like to be around people and are pretty chatty. They are good at understanding and leading people. They learn best in cooperative sharing. Taylor calls them "Socializers."

7. Intrapersonal Learners prefer to work alone and follow their own interests. They are good at being original and pursuing goals. They learn best working in individualized projects. Taylor calls them "Individuals."

It is not too difficult for students to quickly begin classifying themselves as "like this...but NOT like this." But some are confused because they find their style preference may depend on the subject. One student explains, "I think I learn well with visual representations in math, but I really prefer to learn by storytelling in social studies." Laura tells them that the human mind is too complex to be neatly categorized by any system. However, information on MI should provide them with a sense of their general learning preference and suggestions on how to cope when it may be in conflict with a teacher's instructional style.

ENGAGING STUDENTS

To exemplify the teamwork approach for applying MI to a task, Laura presents the following problem to students:

5 oranges and 10 apples cost $3.50, and 1 orange and 1 apple cost 50¢. How much does an orange cost?

She asks: "Think of the ways different learners might approach this problem from a MI perspective." With her help, students suggest that the linguistic and interpersonal learners may prefer to begin discussing strategies; logical learners may begin working with the numbers first; spatial learners may prefer to draw a picture as shown in Figure 4.1; kinesthetic learners may choose to use a manipulative to represent the apples and oranges and arrange them as in Figure 4.1; intrapersonal learners may ask for time to think alone before engaging in group discussion.

FIGURE 4.1 APPLES AND ORANGES

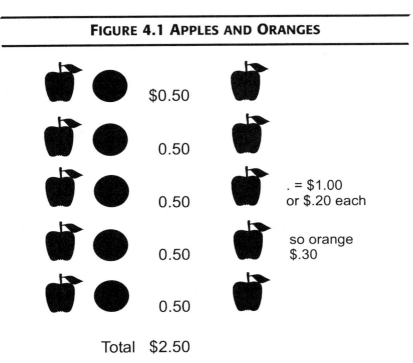

"But what about the musical learner," a student asks. Students learn that not all styles can be easily represented all the time in all problems. However, Laura stresses that now that they have ideas about their preferred styles, they are empowered to think of ways to make learning and studying meaningful. Thus, having an idea of their preferred learning style should not be used as an excuse not to learn, but as an opportunity to learn more about how they learn. For homework, she has students write about the learning style they think best represents their way of tackling problems and to test their guess by doing Investigation 1 shown in Figure 4.2.

FIGURE 4.2 THE STAIRCASE PROBLEM

Investigation 1—The staircase problem: Given a staircase with 5 steps composed of 15 blocks, create a presentation for three ways to find the number of blocks required for a staircase with 100 steps. Use whatever tools will help you: calculators, computer spreadsheet, or manipulatives such as rainbow cubes.

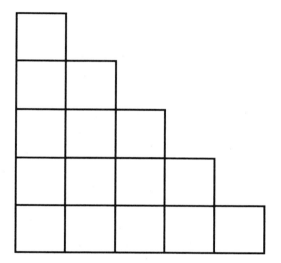

The next day, Laura collects the test so that she can gain as much information about her students' learning style. For the next two classes, she uses this information to assign students to teams to work on an investigation. Whenever possible, she creates teams of seven students, one representing each intelligence. For groups of six, she asks the intrapersonal and interpersonal students to choose one of those intelligences; for teams of five, the linguistic and musical are also combined. Students are asked to play the role of a type of intelligence. The first time they try, decision making is hard, even though Laura has clearly defined the roles as the following:

◆ The linguistic and musical students are in charge of determining the presentation of the problem.

◆ The logical/mathematical, spatial, and kinesthetic students are in charge of the three solutions. The students playing these roles begin to solve the problem, while the rest of the team discusses their contributions to the presentation and solution process.

- The interpersonal students are the taskmasters in charge of keeping everyone on schedule and happy. These students often pair up with one of the solution finders to help them work out a good solution. Sometimes they will explain what another student did. Sometimes they may also decide who will be in what roles. Some students interpret this role as a secretary recording what is going on and who's doing what.

- The intrapersonal students are in charge of checking the validity of the three solutions.

Laura circulates about the room as students work. She observes and notes quite a bit of interesting information about her students. She uses a checklist chart with the MI on the left, and she records which roles students choose on the right. This information will also help her in assigning students' roles throughout the year. She observes that students do not always stay true to the character of the role (which she thinks is fine), but she still holds them accountable for fulfilling the duties of the role. For example, if the solution person cannot solve it by drawing pictures, he/she may ask for help from others but is still in charge of recording the solution.

As she expected, students approach the problem from different perspectives. Some use numerical models, and others use geometric ones; some create tables and search for patterns, and others assume that the staircase is a triangle and apply the area formula to get $(.5)(100 \times 100)$. Still others apply brute force by either adding all the numbers on a calculator or by taping many pieces of graph paper together to facilitate their count. Not surprisingly, lots of mistakes occur.

In all cases, Laura challenges students to try and find a general rule or formula that may justify their answer. Because the scoring rubric (Figure 4.3) assigns points for varying degrees of success with the problem and the use of more than one approach, it plays an important part in encouraging students to pursue not only a solution but also to go the "extra mile" by continuing to think about the problem after they have found an answer. For example, Laura observes that some groups present multiple ways for getting examples to generate patterns. One group pleasantly surprised her by including the process of revising their work in their presentation. That group noticed that the sum of numbers from 1 to 10 is 55 and wrote, "but 10 x 55 will not give you the answer because 11 to 20 = 155, and so on." She writes: "Wow! By doing so, they actually realize that the thought process is worthy enough to be documented!" Other groups include prescriptive strategies to help decide when a particular method works best. For example, contrary to what some students did, one group recommended using a calculator to find the sum manually only if the number of steps is *small*. Still others use the computer to write a simple basic program to recursively add the numbers, or to copy and paste squares to draw a staircase.

FIGURE 4.3 ASSESSMENT SCORING RUBRIC

5 Exemplary Response

- Successful strategy, complete, with clear explanations.
- Shows understanding of the mathematical concepts and procedures.
- Satisfies all essential conditions of the problem
- Goes the "extra mile" by going beyond what is asked for in some unique way.

4 Complete Response

- Potentially successful strategy with clear explanations; may have minor miscalculation.
- Shows understanding of most of mathematical concepts and procedures.
- Satisfies all essential conditions of the problem.

3 Reasonably Complete Response

- Good start on a strategy; may lack detail in explanations.
- Shows understanding of most of the mathematical concepts and procedures.
- Satisfies some essential conditions of the problem.

2 Partial Response

- Start of a strategy; explanation may be unclear or lack detail.
- Shows some understanding of most of the mathematical concepts and procedures.
- Satisfies some essential conditions of the problem.

1 Inadequate Response

- Incomplete; explanation is insufficient or not understandable.
- Shows little understanding of the mathematical concepts and procedures.
- Fails to address essential conditions of the problem.

She observes how well the students understand the concepts or follow through on their roles during each group's presentation of strategies and solutions. Depending on solutions presented, Laura follows with a second investigation consisting of a related example to enrich and highlight possible approaches to the problem.

INVESTIGATION 2

For each of the three examples to follow:

1.   Read each team's approach to the staircase problem.

2.   Describe the method in words.

3.   Try to draw a model to represent the method.

4.   Use the method to compute the number of blocks in 50 steps. Justify your thinking.

5.   Try to use a pattern to discover a formula for the method.

◆ Example 1

One team used the following method for finding the number of blocks in a 100-step staircase.

- $(1 + 100) = 101$
- $(2 + 99) = 101$
- $(3 + 98) = 101$
- $(4 + 97) = 101$

The team continued in a similar way to get $(50 + 51) = 101$ and concluded that $(101) \times 50 = 5050$. Thus a 100-step staircase has 5050 blocks.

◆ Example 2

Another team of represented the problem as shown in Figure 4.4.

**FIGURE 4.4 ONE GROUP'S APPROACH
TO THE STAIRCASE PROBLEM**

We created a rectangle by fitting two of the staircases together. For this rectangle we get (5)(6) = 30 but number of blocks is only 15 because we doubled it.

For 100 stairs we get the area of the rectangle is 100 × 101 = 10100, and the number of blocks is ½ of that: 10100 / 2 = 5050

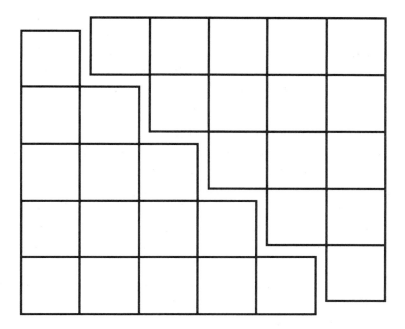

In the review of the first two examples, Laura asks students to explain *what* the numbers represent and *why* the method works. This requirement forces students to think about the connections between the models and the numbers, a very crucial step towards algebraic thinking. In Example 2, students are not quick to see that the numbers connect to the length and width of a rectangle. To help them understand why N steps has (N)(N + 1)/2 blocks, she has students complete a table for simpler rectangles, count the number of blocks, compare the length and width of the sides, and then to try the area formula to see what modification is necessary to yield the number of blocks. After students do so for four or five models, she focuses attention on the dimensions and has students represent the dimensions using N. Next she has them apply the area formula and finally compare the formula's answer to the actual work done with the simpler models. She asks students to determine what should be done to the formula to get the answer (see Figures 4.5 and 4.6). Interestingly, even after having gone through this process, some students still think that the answer for 50 steps results from simply taking the answer for 100 steps and dividing by 2. When this happens, Laura tells them to test their theories, write about their strategies and rationales in their journals and state how their ideas connect to what happened in class.

In Example 3, she presents yet another application, but this time from an historical approach.

♦ Example 3

Here's a neat true historical story: To keep students quiet while the teacher completed work at her desk, young Karl Gauss's (1777–1855) class was given the busy work of finding the sum of all the consecutive numbers from 1 to 100. Carl stunned his teacher by quickly doing it in his head. How do you think he did it?

Students never fail to seriously suggest that Gauss probably used a calculator! Part of next day's homework requires students to develop and test their conjectures, and to find the date for when a hand-held calculator was invented.

Students' results show that most do not readily connect this problem to the staircase investigation. But Laura expects this and writes, "I want to get a picture of the student who does make connections and the student who does not, and then use this information to help them continue to grow at their level of understanding."

It is evident that Laura is clear about what she wishes to assess in her students and that the resulting information from her students' performances informs how and what she teaches next.

DISCUSSION BETWEEN COLLEAGUES

In general, what does a typical day in your class look like?

A typical day—this is middle school so not much is typical, but the following are some general approaches I use:

- We discover some things through questions like: How would you add 100 numbers? How would you find the area? How would you compare video stores for the best price?

FIGURE 4.5 TABLE FOR DISCOVERING FORMULA FOR THE STAIRCASE PROBLEM (ANSWERS ARE IN ITALICS)

Directions: Complete the table. The first two rows are already done for you.

# Steps	Stairs	Rectangle	Width	Length	Area of rectangle	Number of blocks	Compare number of blocks to area of rectangle.
1			1	1	1	1	$1 = 1$
2			2	3	6	3	$3 = 6/2$
3			3	4	12	6	$6 = 12/2$
4			4	5	20	10	$20 = 10/2$
5			5	6	30	15	$15 = 30/2$
n	...n	...?	N	$N+1$	$N(N+1)$	Call it B	$B = N(N+1)/2$

Verify your formula for N = 1 to 5.

- I have them memorize some things: "Look how product of con-secutive numbers are represented." "Oh, just memorize 1/4 = .25 because you'll see it again and you should not have to reinvent this every time. You deserve to have things easy." "Get to know five mathematicians because they merit it. How many are women?" "Memorize as many digits of pi as you can, because, you can."

- We tell some things: "How do they find deer populations?" "How can we represent large numbers?"

- We try whatever works to understand some things: "Try it on the spreadsheet." "Use a table, graph it, draw a picture." "Turn it around 10 times and try it again."

All the while it is important to note that we make *decisions* about all things. It is also each student's job to contribute ideas to the class and discuss his/her choice of method. I often redirect class by asking kids to paraphrase and use statements like "Marie has contributed the idea that...." "Is there anyone...?" "What would you like to contrib-ute?" "If we look at the ideas Bernadine's and Joseph's team contrib-uted...." "I can't hear what Justin is trying to contribute." To the ex-tent students often raise their hands and start their sentences with "I would like to contribute..." these "contributions" are often what stu-dents can add to their written responses to get a 5-point response on our rubric.

What classroom procedures and arrangement foster your students' learning?

On classroom procedures, students enter class and have assignments on an assignment section of the board. They are given a math note-book for each unit and complete a table of contents of their notebook with references to each numbered page. Notebooks include activities, notes, journals, homework, and so on, and are graded. Students have the option to organize the notebook in a way that will be useful for them. Students have bins of materials and computers available to them at any time. It is their choice to get up and get rainbow cubes, number cubes, multilink cubes, pattern blocks, string, scissors, rulers, and/or tracers whenever they think it will help their pursuit of a mathematical question. The idea is for students to easily access mate-rial and information. Sometimes the seats are in a large circle and other times in groups. The seats are so arranged so that groups, part-ners, and teams can change every two or three weeks.

Comment more on how your student's adhered to the roles connected to their learning styles.

> Students do not always adhere to their roles while they worked in their groups because some mathematical problems lend themselves to various roles. For example, musical talent was not easily applied to the staircase problem, unless you like "Stairway to Heaven" 45 times a day. However, students made those decisions and contributed in other ways, like helping one another fulfill his/her role. Again the idea of roles is to allow students to value individual contributions and try to view things from many angles.

How do you use information on learning styles in future lessons?

> Information of learning styles can help me make decisions about forming partners or types of groups to cooperatively work on a project or task. It can also help me make decisions on how to present a problem so that more students can be successful.

What is your general approach to assessment? Do you use traditional assessment? If so, to assess what kinds of skills or concepts?

> In general, I assess students through group presentation, participation with partner/class, journal reflections, and individual and group testing. Students also use self-evaluation on journal entries, presentations (formal and informal), and homework using the scoring rubric. I do whatever I can to get information on student thinking to share with students, parents, and any other interested person. I look to the objectives and outcomes of the Connected Math Project units and our units as guidance. We take individual tests to view problem solving, application, and use of skills or concepts. We take group tests to learn test-taking strategies and assess ability to use team skills. We do group projects, partner projects, and individual projects to apply concepts and skills, make mathematical connections, and make real life connections. There are student-evaluated assignments, teacher evaluated assignments, and group evaluated assignments. We often create a system like the 5-point response (scoring rubric) to assess these assignments. Since so much of the information on these assignments is in written form, I turn for help to the English teacher. I learned that in my students' English class, they use FCAs (focus correction areas) to help assess writing. I think about FCAs when I assess students in math. For example, one can see lots of math in the staircase problem, but I try to focus assessment and comments on strategy and process,

not just calculation and answers. Those FCAs change for other problems.

I send home progress reports every four weeks and give students reports about every two weeks, and students have access to their grades at any time. The reports are organized by exams, group tests, homework, notebooks, projects, team projects, and other, so students can see strengths and weaknesses in these areas of performance. I use a grade program to organize these grades, and I send home postcards for excellence in work or behavior, make celebration calls, contribute to the team newspaper, and write math letters for each unit to better inform students and parents of our endeavor. In order to assess learning, we often seek answers to the question: "Is it not working because of lack of effort or lack of understanding?" With the help of my team of teachers, we try to create a "no excuses" environment. Students are held accountable for their success. My students will study and will be held accountable. If my students don't do homework, I call home. If my students forget their calculator, they come to tutoring and I sometimes teach them by rote the square root algorithm, long division, and so forth. Any single one of these things certainly has faults, but together it gives us a lot of interventions to help kids succeed.

Laura, this seems like a huge amount of work. What support do you get to do it all?

I have the help of my interdisciplinary team consisting of teachers in the areas of math, science, English, social studies, and art/music; a cooperative math department; and district in-servicing for curriculum and technology. With that support, I can communicate with my students in a number of ways.

Where do you search for rich problems?

I use the NSF-funded middle grades curriculum, called Connected Mathematics Project (CMP), which is available through Dale Seymour (phone: 800-872-1100), and some interdisciplinary projects. In addition, I attend at least one NCTM conference every year, and read the *Mathematics Teacher* and *Teaching in the Middle School* journals religiously. We have one math meeting per six-day cycle to discuss ideas with math teachers and four team meetings to discuss ideas with other teachers. I coach MATHCOUNTS and sit down and do the problems with my team. I have math activities and chess club, and I play math games with my students. I take courses, join curriculum-writing committees, and use my own problems sometimes to generate problems. However, my richest problems come straight from my students. For example, in the computer lab one of my students asked, "How could we show the population densities we computed on a map of the United States?" Students began the pursuit and were making dots, using shades, and resizing states to represent population density instead of square miles, and of course writing the numbers on the map, too.

Can you suggest ways for extending this lesson to students in special populations?

These investigations are adaptable using models, and visual cues and partners. For my ESL students I find I often have to act things out or create a model. The staircase can easily be built for students and they can be led to organize sums in a table. One of my students helped Enrique understand the problem by numbering the blocks, 1 2, 3, 4, etc., and using his fingers to take steps up the staircase. He created several staircases and asked, "How many steps?" Once the definition of steps was understood, the problem became accessible to Enrique.

Any further tips for readers?

I have a word of caution for the Gauss problem given in Example 3. Most teachers might prefer to use it as a nice historical introduction to the staircase problem. The first time I tried to do so, there was one student who thought to do it like the staircase problem, and 123 students who merely applied brute force by trying to add the numbers. Gauss's story mentioned addition, so students did just that—added. Next time around, I saved Gauss's story for last. Since the staircase form of the problem merely asks for the number of blocks, students presented more than just addition as a strategy. Keeping Gauss's story last works better because I can then see which students can connect what we did in class to the story without inadvertently stifling creativity.

Tell us about your school's minority participation in your classes. Do you notice any differences in their performance in this lesson?

The largest barriers to high performance in my math class, especially since the adoption of CMP, has been language because of the amounts of reading and writing. In addition to any adaptations I make, our school has an ESL teacher, a learning support teacher, and a communication teacher to help adapt materials for students with reading or writing difficulties. I find any of the adaptations requiring more models, visual cues, and partners are helpful for every student and do not limit the use of these ideas to a select few. We thrive because of our differences, not in spite of them.

COMMENTARY

Laura's dual application of a problem to gain insights not only on her students' problem-solving skills but also on their learning styles underlies how mathematics can be made accessible to all students. The different approaches taken by her students also reveal how necessary it is for students to have multiple representations of an idea in order to build solid conceptual understanding.

Her investigations have characteristics important to reaching all students. They are conducive to students' use of multiple ways to express their knowledge, thus tapping into students' preferred mode for learning. The fact that students can use manipulatives, and are encouraged to express their formulas in words, numbers, or algebraic symbols, invites students to move from a concrete example to others requiring greater and greater degrees of abstraction (algebra, representation). Letting students know that their ideas and learning preferences are valued increases the likelihood that lower-achieving students will feel safe enough to share their own ideas, while encouraging higher-achieving students to take greater risks (Equity).

I was most impressed with Laura's use of knowledge of the learning styles of her students to guide her instructions. In a class of students with varying abilities, teachers adhering to the *Standards* typically create lessons with a variety of activities to accommodate different learning styles. But what does a teacher do when she/he knows the actual *specifics* of the learning preferences of each student in the class? How we use that knowledge may differ. Some may continue teaching in same manner and tell students that it is *the student's* responsibility to use that knowledge to enhance learning. This warns the student that if he/she has a teacher who lectures often, but he/she prefers to work hands-on, then some learning problems may ensue and it becomes the student's responsibility to apply strategies for coping and succeeding in the course. Others, like Laura, may decide to make a conscious effort to help students develop the coping strategies. When she says, "I try to get them to see things from many angles," or when she purposely varies the roles they play in teams, she is helping students develop strengths in a different learning style. This practice is crucial if students are to succeed in becoming flexible thinkers with the ability to process information from various sources or modes of delivery (curriculum, teaching, assessment, problem solving, communication connection, reasoning, and representation).

For readers interested in knowing about students' learning preferences, Laura's references should prove useful. There is also a hard-copy inventory by

Johnston and Dainton where students answer questions and then assess them quickly along the dimensions of learning styles (Sage Publications, 2455 Teller Road, Thousand Oaks, CA 91320, phone: (805) 499-0721, Web: http://www.order@corwin.sagepub.com). Suzanne Miller has a Web page with 32 learning style questions for students to answer on line (http://silcon.com/~scmiller/lsweb/dvclearn.htm). The answers are processed immediately to give a student an idea of his/her styles together with excellent suggestions on how to best learn under adverse conditions. Although intended for college students, the inventory can be helpful to middle or high school students who are given the kind of help Laura provided for her students. These resources, and Laura's approach for helping students understand particular learning styles, are crucial to helping students take responsibility for their own learning while simultaneously reducing the levels of mathematics anxiety.

CONTACT

Laura Mullen
Mt. Nittany Middle School
656 Brandywine Drive
State College, PA 16801
Phone: (814) 466-5165
E-mail: Ljmll@scasd.k12.pa.us

UNIT OVERVIEW:
PROBLEM SOLVING AND LEARNING STYLES

Aim: How can a mathematical model help us to think about a problem in many ways?

Objectives: Students will apply multiple strategies requiring different representations for solving a problem.

Grade Levels: 7–9

Source: Original

Number of 45-minute periods: 2–3

Mathematics Principles and Standards Assessed:

- Principles for equity, curriculum, teaching, learning, assessment, technology

- Mathematics as problem solving, communication, connection, reasoning, representation

- Number and operations

- Algebra

- Measurement

- Geometry

Prerequisites: Students can perform basic operations and apply area formulas for rectangles and triangles.

Mathematical Concepts: Students develop, explain, and apply strategies for adding consecutive integers through numerical, tabular, algebraic, and geometric representations.

Materials and Tools:

- A description of the seven learning styles

- A statement of the staircase problem

- Assessment Rubric (Figure 4.3)

- Spreadsheet program

- Rainbow cubes or other cubes that lock unto each other

Management Procedures:

- Discuss the seven learning styles.

- Ask each student to pick the style that best represent him or her.

- Present the staircase problem.

- Assign students to groups of four and describe their roles according to their learning styles.

- Have students do the problem and present their approaches.

Assessment:

- Assign students to complete an investigation.

- Note the processes they apply and how they communicate their ideas.

- Use an assessment rubric.

5

MADELINE LANDRUM: MODELING REAL-WORLD PROBLEMS WITH MULTISTEP INEQUALITIES

I like to introduce my lessons with an academic challenge by choosing problems that relate to the lesson and simultaneously challenge the students. Such problems allow the students an opportunity to "discover" some math concepts on their own. While the challenges usually create a state of frustration for students, students have come to accept this state as precursor to great mathematical discourse that will eventually guide them through the major parts of the day's lesson.

> Madeline Landrum,
> Lusher Extension Middle
> School, New Orleans, LA

Madeline's students enter her class and immediately sit around tables that are arranged for group work. Students normally sit in pairs, but for this lesson she tells them to sit in any group of three. "I thought three in the group would promote more discussion and brainstorming," she says. For this lesson on inequalities, she chooses a challenge from the text that reads:

> Artists often use the golden rectangle because it is considered to be pleasing to the eye. The length of a golden rectangle is about 1.62 times its width. Suppose you are making a picture frame in the shape of a golden rectangle. You have a 46-inch piece of wood. What are the length and width of the largest frame you can make? Round your answers to the nearest tenth of an inch (Prentice Hall 1997, 193).

Included with the problem is a picture of the Parthenon, the ancient Greek temple that was designed so that its dimensions formed a golden rectangle. As the students begin tackling the problem, she observes different approaches and areas of misunderstanding arising. As examples, one group uses trial and error, and another uses the perimeter formula and substitutes 1.62W for the length; In one group, students have no idea what to do with the "46 inches long" information given about the wood because the problem makes no mention of the words length or width. After questioning the students in one group, she realizes that this group students doesn't have a clear picture of what a 46-inch piece of wood looks like or what they are to do with it.

It is interesting to hear the initial brainstorming about the problem and see how Madeline facilitates discussions as she walks around. Her interaction with a group that couldn't start the problem follows:

Magan: Ms. Landrum, if the piece of wood looks like this (she uses hands to form a rectangle), why don't they give us length or something?

Madeline: No, No. It's just one piece of wood that is 46 inches long. You are going to cut that up so that the rectangle you make...so that the frame you make... fits the requirement that the length is 1.62 times the width. But, the wood is just 46 inches long.

S1: Oh...so we just have to...like...cut it into four different pieces to do that?

Madeline: Right!

As Madeline leaves, she hears Magan explaining to others in her group, and they then begin the planning process. Another group has progressed from understanding the problem to making a plan, but is having difficulty translating the key ideas into algebra. The following discussion reflects how students do not easily understand the variable concept.

Tony: OK. It's going to be 6.2 times x or something like that because you have to use a variable.

Bernadine: Why don't we try a big number first to see how much....

Tony: It's 46 inches, right? If it's 46 inches, then that means it's 46 inches long— right?

Sergio: I'm not sure but I think it's 1 point 62 times, or maybe, plus, x equals....

Bernadine: 46!

Sergio: Yeah, 46.

Bernadine: I think you divide it by 2 times its width, because the rectangle is twice its width long.

Sergio: So it would be 1.62 times…2? Then it would be like, x, or something like that?

Tony: Look at the Parthenon picture—It would be 1.62x, times 2.

Bernadine: So wait…1.62x, twice, times x squared = 46?

As Madeline approaches this group, she demonstrates how to redirect students thinking through questioning. When Bernadine asks her if the group has set up the equation correctly, the following dialogue ensues:

Madeline: What are you trying to find?

Bernadine: Wouldn't you use $2L + 2W$ which would give you P?

Madeline: And what is that?

Bernadine: The perimeter…which is… 46? Is that right?

Madeline: Ask others in your group.

This quick interaction is enough to get the group to correctly apply the perimeter formula. Another group does not understand how to represent the length in terms of the variable. Madeline's line of questions shows ways to help students gain increased understanding on the use of variables: "What are you trying to find for the rectangle? What does your x represent? What is the length? What does the 46 represent in terms of this? Let's go back to the problem. You need to write down what your width represents….OK, so you think the width is the variable—then what will the length be equal to?…" As this group begins to apply the perimeter formula, she moves on.

Madeline walks around to be sure that the groups have reached an answer and thus asks each group to check answers. Helen calls her over and says that she has checked her work twice with a calculator and says, "but when I check, I'm over the 46 inches yet I know my answers are correct." Madeline replies, "But is that possible?" She walks over to another group and after a few more minutes of grappling, calls the class together and has one group go up to the board to set up equations:

Let W = width and $1.62W$ = length

$2L + 2W = 46$	$L = (1.62W)$
$2(1.62W) + 2W = 46$	$L = (1.62)(8.8)$
$3.24W + 2W = 46$	$L = 14.256$
$5.24W = 4$	$L = 14.3$
$W = 8.8$ and $L = 14.3$	

It's interesting to see how this group partitions the task of reporting. While one student writes on the board, the other two take turns explaining what is being written and why. It clear that this group had all its members actively involved in determining the solution. Students then ask questions of the group members, who again take turns explaining their reasoning. A student challenges the group by saying, "But when you check the answer, you get more than 46." The members of the group check the results and show that they get 46.2, but admit that they don't see any errors. Madeline calls on Pablo, who had taken an alternative approach to the problem, to show his work at the board. Instead of an equation, Pablo writes the inequality, $2(8.78) + 2(14.22) \le 46$. When asked why, he explains "you want the largest dimensions possible without using more than 46 inches." He also explains that he rounded the width (8.77) to 8.8, but that he rounded the length (14.25) to 14.2 so as not to exceed the 46 inches. The class agrees with his reasoning, but Liz says, "If you use those signs and those dimensions, then Pablo's inequality is more than 46, so that is still not true. He'd have to put: $2W + 2(1.62W) \le 46$. Madeline looks at Liz's work and notices that the numbers she substitutes for length and width are rounded to the tenth place. She tells Liz, "Compare his numbers to yours and then check his work again." A more careful check satisfies Liz's group, and Madeline asks the class to reflect on why the other inequalities did not work, for homework.

For the next activity, she tells the class that the Greeks not only used the golden rectangle in their architecture, but they also believed that some human body proportions were "divinely proportional" provided they fit the characteristics of the golden rectangle. She instructs students to measure two distances, the first being the measure of the distance from the head to the navel, and the second being the measure of the distance from the navel to the feet. She tells the class, "Use the first result for the numerator of the fraction, and the second for the denominator. Determine how close the result is to 1.62 or to having a 'divinely proportional body.'"

With tape measures and calculators in hand, students enthusiastically set out to measure each other to find the "perfect body." You could hear them saying to each other, "Take your shoes off; stand up straight. This doesn't seem right; let's do it again; divide that with the calculator." After about 15 minutes, students move around and happily share their measurements with other groups. One group reporter, Laura, gives a quick summary: "While Lisa has a divinely proportional body, I am the least divine person in my group!" Laura is very tall, and is disappointed in her ratio because it is over 1.7. Madeline asks students to think of why her ratio might be higher than others. Kendra suggests that "it is probably because her legs are long, so that her torso probably has yet to catch up with them." Others agree.

Madeline assigns students to practice solving multistep inequalities from their text. Again, she walks around and monitors the students' progress as they work in groups. For homework, she assigns additional problems from the text and tells students to gather measurements to compute the divine ratio for one member in their family.

The next day, Madeline begins the lesson by asking students to explain the discrepancies in the inequalities from the previous day's lesson on the piece of wood. Nicette suggests, "Well, in both of Pablo's equations he had numbers in the hundredths but we rounded off our numbers to the nearest tenth. Maybe that's the problem." Students return to their calculations and verify that Joseph and Nicette are indeed correct. Joseph chimes in, "Sure! We shouldn't have rounded off until we reached the end!" "Does that make sense?" asks Madeline as she looks carefully at students' faces for signs of confusion. As students nod "yes," Madeline distributes an overhead transparency with a table on it for students to enter the family data gathered for homework and the corresponding divine ratio calculation. She asks, "Do you think that most people will have that characteristic? After the class has entered the results, I will make you a copy for you to analyze on a spreadsheet." While the table circulates, Madeline reviews the assigned homework from the text and then proceeds to have the class analyze the data gathered on a family member's divine ratio calculation.

DISCUSSION BETWEEN COLLEAGUES

What made you decide to teach mathematics?

> I have always enjoyed math because I had success with it. My dad was also a math teacher. However, it was my methods course instructor who influenced me to consider math. She took me aside one day and said that she thought I would enjoy continuing with math so why not become a math major—I did—it was one of the best decisions I ever made!

Tells us a bit about your school.

> Although some Lusher students are selected based on their good academic record, they come with a wide range of economic backgrounds. We make sure that all students identified by the social workers as being "in need" can get the necessary school supplies, go on field trips, or even receive spending money for our crawfish boil. Donations from our business partners help make this possible. Lusher has an art focus so that teachers try to incorporate art as a motivating topic throughout the disciplines whenever possible. Teacher schedules are arranged so that each grade level has a common planning period for sharing and organizing instruction once a week.

What is your philosophy for teaching mathematics?

> I believe that NCTM is right: All students can learn mathematics—maybe at different rates or different levels, but they can all learn it. I try to help students become comfortable with mathematics because so many come to me afraid of it. By giving students time to work with a concept or through use of different techniques, I find that I slowly break down the walls that hinder their learning.

How do you generally assess students?

> I use both traditional and alternative ways to assess students. For example, I find test scores, quizzes, and grades for notebook useful. I also have students do both individual and group projects and allow students to show improvement on any score that is equivalent to a low grade.

What resources do you use other than the text for instruction?

> I use calculators, computers, the Internet, and manipulatives. The third lesson in this unit, for example, uses the art of Da Vinci's Vitruvian Man to investigate ratio of arm span to arm length, which I downloaded from the Internet. (http://forum.swarthmore.edu)

Middle grade students are so body-sensitive. Do you find that the data analyzed is enough to show that it's OK not to be divinely proportioned?

> Yes, because I discuss this issue at length after the second lesson because the data do support this. This is the primary reason I assigned students to collect data on family members. In the past, I have had them collect data on teachers, too.

What would you keep or change next time you teach this lesson?

> I liked both activities, and I will use them again. Before this lesson, the students had solved some multistep equations. I found the academic challenge to be just what I needed to introduce the lesson in a way that demonstrated to students the close parallel between solving equations and inequalities. Students had difficulty seeing the shape of the 46-inch piece of wood. Maybe a general class discussion clarifying the given information might help. Students also began rounding off their decimals too early in the process. I left it as a challenge for students to determine why their equalities did not work but other times I may just give them a warning and move on.

Do all students take algebra? If not, how does Lusher determine who does?

> Lusher places 8th graders in an algebra class for high school credit or in a prealgebra course for graduation credit only. Both courses are taught from a reformed-based perspective and use the same book, but we spend more time on concept and skills development in the prealgebra. We use IOWA test scores, averages in math classes, scores on the basic skills test, and teacher recommendations to decide who takes an algebra course. We find these measures to be good predictors of success. However, there are times when we allow adjustments after placement for a prealgebra student who, for example, demonstrates good potential to succeed in algebra or for an algebra students who is struggling. In both cases we offer the students the option to switch.

COMMENTARY

Madeline's lesson provides opportunities for students to use algebra to model and solve problems—a most difficult concept for many students to grasp even with each other's help. From his research on students' attempts to apply variables while solving nonroutine problems, Reeves (2000) writes, "Students will not automatically learn to use variables even after hearing their classmates use them as shortcuts. The use of variables will have to be encouraged by the teacher if that outcome is a goal of an algebraic-thinking strand." (p. 401). As she listens to her students grapple with applying a variable to a problem, Madeline applies Reeves's advice through questions that reflect a sound method for helping students focus on the essential aspects for applying variables. Her questions reveal to students that they too can get fruitful answers if they ask themselves and each other similar questions—questions that focus on understanding the variable's role in the problems and which Madeline asks over and over again as she lays the foundation for students' deeper understanding.

Madeline's lesson also connects geometric ideas to art and everyday life (geometry and spatial sense). The difficulty students have understanding what to do with the information given on the 46-inch-long wood illustrates how necessary it is to give problems requiring analysis of relationships between two-dimensional and three-dimensional figures. Such problems help students learn that real-life problems may not have the key words that readily connect to algorithms they know.

In their article on algebra, Schappelle and Phillip (1999) write, "The way in which a teacher views algebra has important implications for their instructional goals. A teacher who believes that algebra is primarily about manipulating symbols will teach algebra differently from someone who sees algebra as a language for generalizing arithmetic" (315). Madeline's approach clearly demonstrates that she believes in the latter.

Her homework requiring that students find measurements of a family member has an important outcome extending beyond students' practice of skills. It is a good way for informing families about the content students are learning through a fun and meaningful activity, which may help families make some sense of mathematics reform. Such assignments often get parents to say "Wow! As a student, I never had this much fun doing math!" This sort of information sharing between classroom and home is important in assuaging some parents' fears of mathematics reform being synonymous to a watered-down-fun curriculum or having little connection to traditional basic skills. Interestingly, Madeline's outreach in this domain extends even further: The Linton Professional Development Corporation (2000) produced a video focusing on the *Standards* to help educators and community members better understand the application of the *Standards* to the classroom (phone: (800) 566-6500, e-mail: www. videojournal.com). Madeline, as one of the teachers featured, can be viewed teaching this lesson.

The multiple assessment approach Lusher applies to determining which students take algebra for high school credit is commendable and shows great concern for equity—this includes student's preference to stay in the course even if he/she is failing! In their article reflecting on the appropriate role and importance of algebra in the school mathematics curriculum, Strong and Cobb (NCTM 2000) pose questions for which, in their minds, "answers to these questions are fundamentally related to issues of equity, expectations, and effectiveness":

- ◆ If not Algebra, then what?
- ◆ If not for all children then for whom?
- ◆ If not for all schools then in which ones?
- ◆ If not now, then when?

Lusher's courses in both algebra and prealgebra do an excellent job of addressing the issues of equity, expectations, and effectiveness: All students take algebra or prealgebra; the courses use the same text and are taught from a reformed-based perspective; and the courses differ in pacing, traditional basic skills focused, and attainment of high school credit. In effect, students in the algebra courses receive instruction similar to what Strong and Cobb call "Bi-Mathematics, which is similar to bilingual instruction in that it produces students who are proficient in two languages—skills-based and standards-based —of mathematics" (3). Students passing these courses will have a strong algebraic background, which will help them to succeed in high school algebra.

CONTACT

Madeline Landrum
Lusher Extension Middle School
719 South Carrollton Avenue
New Orleans, LA 70118
School Phone: (504) 862-5114 Fax: (504) 862-5167
E-mail: MLLNDRM@aol.com

UNIT OVERVIEW:
SOLVING MULTISTEP INEQUALITIES

Aim: How can inequalities help in solving real-world problems?

Objectives: Students will model and solve a real-world problem from geometry using multistep inequalities.

Grade Levels: 7–8 algebra

Number of 90-minute periods: 2

Source: *Algebra: Tools for a Changing World* (Prentice Hall: 1997), p. 193, Example 42.

Mathematics Principles and Standards Assessed:

- ◆ Principles for equity, curriculum, teaching, learning, assessment, technology

- ◆ Mathematics as problem solving, communication, connection, reasoning, representation

- ◆ Numbers and operations

- ◆ Patterns, functions, and algebra

- ◆ Measurement

- ◆ Geometry and spatial sense

Prerequisites:

- ◆ Solving one-step inequalities

- ◆ Concept of ratio and proportions

- ◆ Ruler measurements

Mathematical Concepts: Students apply the perimeter formula for rectangles to solve a real-world problem using multistep algebraic inequalities.

Materials and Tools:

- ◆ Tape measure

- ◆ Calculator

Management Procedures:

- ◆ Assign students to groups of 3–4 to work on finding golden proportion dimensions for a rectangle whose perimeter cannot exceed 46 inches.

- ◆ Have students share results.

◆ Have students measure length of body parts to seek the 1.62 divine ratio stemming from length of navel to foot, divided by length of head to navel.

◆ Assign multistep inequalities for practice.

Assessment: Circulate to observe and question students' work. Check multistep problems assigned for homework and check students' search for divine ratio in a family member's measurements.

6

KIMBERLY McREYNOLDS: THE GEOMETRY OF KITES

I personally believe that students should have as much hands-on experiences in math as much as possible. I don't like to tell my students the "rule." I like them to discover it themselves. To do so, I write math curriculum units, along with integrated curriculum units, which focus around one theme, concept, or idea. My unit, "Kites and Other Flying Objects!" started around the theme of kites, and I took it from there, researching and ending up with a unit that now includes language arts, math, science, social studies, and art. Tell me an idea, and I can incorporate math into it 98% of the time. I LOVE MATH!!

Kimberly McReynolds,
Mathematics teacher,
Cresthill Middle School,
Highlands Ranch, CO

Kim's quote led me to believe that she probably has had pleasurable and rich experiences with math while in school. Au contraire, for Kim writes: "I was taught in a school where I was given the rule and then had to solve 50 problems using that rule. I absolutely HATED math as I was growing up." She began liking math only after she was hired as a Chapter I math teacher and was informed that the philosophy of the program was based on 90% hands-on explorations in math and only 10% on paper and pencil. "WOW!" she questioned, "Math can actually be taught in ways other than paper and pencil?" She participated in workshop sessions showing her how it could be done, and since then she has been growing in her love and appreciation for mathematics. Her new-found feelings are expressed in enthusiasm for creating challenging and interesting lessons for her students. Her unit on kites consists of 10 lessons and requires between one and two months if taught in its entirety. We will focus on six of those days.

PREPARATIONS

Before beginning the actual activity of making the kites, Kim prepares her supplies and decides on strategies for getting funds to acquire additional materials needed. Her presentation of the goals of the project to her school's Parent and Teacher Organization is generally successful in generating funds. Her next steps are to create clear and explicit directions and to decide on a rubric checklist for students.

Kim begins by having a warm-up activity asking students to solve a riddle or inference on the making of an object.

INFERENCE

Given the following facts about how an object is constructed and some of its characteristics, try to guess the name of the object:

A newspaper is better than a magazine, and on a seashore is a better place than a street. At first, it is better to run than walk. Also, you may have to try several times. It takes some skill, but it's easy to learn. Even young children can enjoy it. Once successful, complications are minimal. Birds seldom get too close. One needs lots of room. Rain soaks in very fast. Too many people doing the same thing can also cause problems. If there are no complications, it can be very peaceful. A rock will serve as an anchor. If things break loose from it, however, you will not get a second chance.

Anonymous

After students have read it, she directs them to quietly discuss with their partner or group what they think the statement is about. After several minutes of quiet discussion, she asks each pair or group to write guesses on the board. A lively discussion ensues as each group tries to defend guesses that generally suggest outdoor sports. Thus far, none of her students have ever guessed that it is about making kites. The next step is to tell students that they are ready to embark on a wonderful unit and to invite them to try and guess what the aim for the unit of study will be using the inference as a guide. If no one guesses that it will be about kites, she gives additional hints until a student guesses it, then has them verify that kite satisfies the inference. To keep note of their ideas, she makes a "KWL" chart which contains three columns: The first column is titled, "What we *know* about kites"; the second column is titled, "What we *want* to learn about kites"; and the third column is titled, "What we *learned* about kites." To sow seeds for further questions and ideas for independent research by the class or individual students, Kim adds an additional column titled, "What we still want to learn about kites." She keeps this chart posted in the classroom throughout unit, so that she and/or her students can continually add to it during each lesson. The students' second task is to do some research on the history of kites, and/or Kim provides this history (see Figure 6.1). As they encounter new terms, Kim requires that they create and sustain a *Dictionary of Geometric and Kite Terms*, which is divided into four sections: Kite Definitions, Geometrical Definitions, Airplane Definitions, and Other Definitions. Students may use words or pictures to define terms, and Kim has a glossary of terms available for students to use if they choose. In keeping with the theme, students create their dictionaries in the shape of the kite of their choosing.

Kim also reserves a section of her bulletin board for the definitions. Choosing a specific section of the dictionary, say, kites, students cut small kites from construction paper, write a term on the front of the kite, put the definition on the back, and attach a colorful tail made of crepe paper. The definitions make pretty displays on the bulletin board and may also be hung from the classroom ceiling. Different categories or sections can be made with different colors of paper and similarly hung from the ceiling throughout the unit. For example, all geometrical terms can be put on blue kites, and so on. Interestingly, Kim also uses them to further students' learning by playing a form of the game "Jeopardy." She divides the class into teams, calls out a definition, and asks a student for the corresponding term. At a later time, she removes the definitions on kites and replaces them with, say, geometrical definitions, and the game begins again. Students thereby painlessly learn the terms and the definitions, while the class periodically enjoys a fresh bulletin board display.

ENGAGING STUDENTS

When students get the directions for making their kites, they are surprised to see that the goal is not to simply make pretty kites. Kim's instructions on required geometric figures quickly reveal to students that the final product will require thought, practice, and time (see Figure 6.2). Students are quick to undertake the challenge because, as one said, "This is fun stuff!"

For the next four or five days, students busily cut, paste, read, talk, giggle, and move about the class to observe and comment on each other's ideas.

(Text continues on page 81.)

FIGURE 6.1 BRIEF HISTORY OF KITES

- The first kite flight was recorded back in China about 200 B.C.

- Kite flying is more than 2000 years old.

- Kites were used in wars. Scary-looking kites or kites that made sound when the wind rushed by them were used to frighten enemies. People flew the barrage kite (looks like a capital letter I) in front of enemy planes to make them crash.

- Benjamin Franklin: In June 1752, Benjamin Franklin discovered electricity when he flew a kite in an electrical storm. Up to that point, people believed that lightning had mystical meaning that struck terror in people's souls; the deadly bolts from heaven had been perceived as the wrath (anger) of the gods. Benjamin Franklin believed that if lightning could be drawn to a kite in a storm, lightning could be safely absorbed by a grounded iron rod attached to a house. He showed how lightning could be controlled. Lightning rods began rising over buildings in Philadelphia, Boston, and New York City, and soon in Europe, too, all because of a kite.

- Alexander Graham Bell: Alexander Graham Bell made very big kites. In 1907, one of his kites lifted a man 200 feet above the ground! He invented the tetrahedron kite, which means "having four sides." It is a kite that consists of sticks arranged in one or more triangular cells. On December 6, 1907, Mr. Bell launched his 208-pound kite over Baddeck Bay, Nova Scotia. It flew 168 feet off the ground for approximately 7 minutes. Dr. Bell believed that the structural strength of the triangle, amplified into a triangular pyramid, would produce a super-strong lightweight framework for a kite.

- Wright Brothers: The Wright Brothers used kites to help them design their plane.

- Others: Lawrence Hargrave (of Australia) decided that his future lay in scientific inquiry, and he soon became interested in artificial flight. His kites (in 1894) lifted him up in the air. His experiments led to air travel, although he was never credited. He had intended to invent the airplane, but instead he invented the cellular or box kite. He never patented his idea, but freely shared it with others. Box kites have been used to rescue pilots who have crashed.

FIGURE 6.2 GUIDELINES FOR KITE CONSTRUCTION

Name _____

Part 1: REQUIREMENTS: You must use a straight edge at all times.

1. Minimum requirements for your Kite:

   a. All lines are straight. They do not have to be congruent.

   b. Two angles. They do not have to be congruent.

2. Minimum design requirements on front side of kite:

   a. 1 quadrilateral

   b. 1 polygon other than a quadrilateral

   c. 4 parallel lines

   d. 4 line segments

   e. 2 rectangles

   f. 2 perpendicular lines

   g. 2 rays

   h. 3–4 points

   i. 2 parallelograms other than rectangles

   j. 2 circles

3. On the back side of your kite, label the following items in pencil:

   a. all angles: number of degrees, and whether each is acute, right, obtuse, and/or congruent.

   b. all shapes: whether polygons, quadrilaterals, etc.

   c. all lines: whether parallel, perpendicular or intersecting, any line segments.

Part 2: PROCEDURES

1. Use notebook paper to brainstorm possible designs for your kite.

2. Decide on a design and check that it meets the above requirements.

3. Use the draft paper given to you to draw a final draft kite.

4. On the back of the draft kite, write your first and last name neatly in a corner of the kite, small, please, so as not to interfere with any labeling.

5. Show teacher your draft kite.

6. Use final paper to draw the final kite.

7. Copy your draft kite onto the final paper.

8. Use colored pencils to completely color the front side of your final kite from edge to edge.

9. Note that the dowels of store-bought kites are the "skeleton" of the kite and are actually on the front of the kite, which is the same side to attach the string. Our dowels and labeling will be on the back so as not to interfere with the design of our kites.

    After all coloring and labeling have been done, glue the dowels on the back of the kite and let them dry overnight. Keep this in mind for turning your kite in on time so that the glue is dried. Keep the rough-draft kite until you are completely finished with the final kite.

10. Glue strands of crepe paper as a tail for your kite.

Due dates: Draft Kite: no later than _____

              Final Kite: no later than _____

They must first make a sketch on notebook paper, followed by a draft on 3-inch x 4-inch paper before they can make their final kite. If they aren't sure what each of the requirements are, they don't ask Kim, because she expects them to use each other as resources and to consult the rubric checklist (Figure 6.3). Kim circulates to answer questions and informally assess students' involvement and areas of difficulties.

Once they have completed their final kite, both Kim and the students complete the rubric checklist to assess the work. Students easily assess whether their kites have met all the requirements from the list. Kim does not expect or allows failures. She holds high expectations for all students and requires that those getting less than a grade of C redo or modify their kites by a given date.

DISCUSSION BETWEEN COLLEAGUES

Kim, after completing this unit, your students must be "flying high" on self-esteem and appreciation for mathematics. In general, what does a typical day in your class look like?

> My students are used to lots of hands-on activities for learning their basics. I generally start class with a warm-up activity to stimulate some thinking. When beginning a new concept, I have students work with a manipulative that helps the concepts to evolve. For example, I am currently developing a unit integrating fractions, decimals, ratio, and percents, where students examine cereal boxes to determine how they are used to communicate important facts about the cereal and the key mathematical concepts. Paper-and-pencil calculations to reinforce and extend the concepts will follow later.

Describe students' reactions or comments as they do the unit.

> Students are very enthusiastic throughout the project. Their comments to me include, "This is math?" and "Great! What comes next?" Both are said with a sense of appreciation for the project.

Elaborate on students' assessment during the project.

> Many of the activities throughout the unit can be forms of assessment. If the students follow the given directions in the lesson, the finished product should work, perform, or fly. The activities students perform are geared toward understanding the concepts covered in the lesson. If a student completes a product that does not work, the student can analyze the product and reevaluate the process. This reevaluation is itself a form of assessment. Students also keep a journal as we work through the unit.

FIGURE 6.3 GEOMETRIC KITES: RUBRIC CHECKLIST

Student's name _____

A. Place a check on the line if your kite has each of the following. A check means *yes*.

1. One polygon	____	6. 2 perpendicular lines	____
2. One quadrilateral	____	7. 2 rays	____
3. 4 parallel lines	____	8. 3 to 4 points	____
4. 4 line segment	____	9. 2 parallelograms	____
5. 2 rectangles	____	10. 2 circles	____

B. Place a check if you labeled the following on back side of the kite. A check means *yes*.

11. All angles ____	Grading Scheme:	
12. All shapes ____	All *Yes* or checks	A
13. All lines ____	1 to 3 *No*	B
Colored neatly: ____	4 to 6 *No*	C
Colored completely: ____	7 to 9 *No*	D
	10 or more *No*	F
	Grade:	_____
	F or D kites need to be redone by:	

Comments: _____

_____

_____

Other forms of assessment that can be used in this unit are informal observations by the teacher. I *constantly* observe students throughout this unit, watching and helping them work through each lesson. I keep anecdotal notes on stick-it note pads by writing the date, student's name, and what I observed about the student that day. This informal observation makes it easy for me to observe several students within a given time period. I sometimes use holistic scoring by assigning the project or performance one overall grade. An example would be, "Joseph's kite was well made; it met all of the requirements and therefore merits a grade of A." Analytical scoring can also be used. I invite students to help me determine key factors to include on a scoring rubric for the final project. I find that their recommendations are generally more demanding than mine. I assign each part of the project a score then add the scores to get a grade.

Do you use traditional pencil-and-paper tests?

I use traditional assessment throughout the units that I teach. I integrate my own questions with those of the classroom textbook.

What are the areas that make the unit worth so many days?

The problem-solving process is naturally modeled in the unit—especially the last stage, requiring that students look back and review the solution. When assessing their final kites, I emphasize that the process of reevaluating their ideas is an important heuristic and that they use heuristics for solving problems in all areas of life. We then review other heuristics available to them such as taking a guess, examining their assumptions, checking that all of the possibilities were tried, constructing a model, listening to other points of view, and "incubation," or relaxing and thinking about something else for a while. The mathematics is never lost as we integrate the real world.

Do students fly their kites?

We do try to fly them as a class activity. Several students who have flown them at home have told me that they look "pretty cool, my kite and all, up in the sky." They do look absolutely *incredible* when finished. To be honest, with the time and effort spent on these geometric kites, most students don't want to fly them. They prefer to keep them safe.

In our communications, you mentioned that the cost of the entire project is approximately $100 to $300, depending on the quality of materials purchased. Comment on how you raise funds.

> In addition to the PTO, I write letters to kite stores in the area and various kite associations. They are extremely helpful in giving donations. Several parents independently help and donate money, too. Money from other school funds may also be available from the school principal. The key is to ask and ask.

How do you find the time to make your own units? What help do you get to make that happen?

> I find the time wherever possible because I really enjoy connecting learning in a novel way. I get help with my ideas from colleagues and students. While some units are just a few days to a week long, others, like Kite, are much longer. I would ultimately like to write and create workshops to help teachers integrate such units throughout the curriculum because the units are powerful tools for helping students see connections and thus gain greater appreciation for mathematics. My students remember important concepts even after exams.

Any tips for teachers?

> For geometric kites that students will really love, six or seven class days are necessary. However, fewer days can be allotted if students are required to take them home to continue the work. Before starting the project, it's important to know which kite students will make and to then set a limit on how much it will cost. Do shop around, because I have found that it makes a big difference on how much I have spent. Note that the guidelines may be adapted to fit the needs of any class.

COMMENTARY

What is very evident in the type of projects Kim chooses is that the child in her is alive, well, and now loves math. This is encouraging. For those who have little hope in reaching students who dislike math and/or have math anxiety, Kim's own negative to positive experience with mathematics shows that there are ways to get kids (and adults) to like mathematics. The key is to find a way to teach it that reduces the anxiety. For Kim, a hands-on approach made mathematics interesting and alive. In a list of practices to reduce mathematics anxiety, adapted from NCTM, Stuart (2000) describes activities or actions that:

- Accommodate different learning styles.

- Create a variety of testing environments.

- Design experiences so that students feel positive about themselves.

- Remove the importance of ego. It should not be a measure of self-worth.

- Emphasize that everyone makes mistakes.

- Make math relevant.

- Empower students by letting them have input into their own evaluations.

- Allow for different social approaches.

- Emphasize the importance of original quality thinking rather than manipulation of formulas.

- Characterize math as being a human endeavor.

Kim's instructional approach and activities put these recommendations into action. Her preference for engaging students in problem-centered learning invites her students to develop their own methods and approaches for viewing and understanding the content under discussion. While she gives them guidelines, students are free to decide what the final product will look like and are responsible for determining whether the outcome of their work reflects the guidelines. The activities she develops has students communicating mathematical ideas through their drawings and final models, defending their viewpoints, using a variety of tools, and gaining a deep understanding of geometric terms and their properties. Such communication occurs when a student feels safe to make mistakes and believes that the teacher and class value his/her thinking. In their study on motivation of 4th- to 6th-grade students to learn fractions, Stipek et al. (1998) found that "the affective climate...turned out to be the most powerful predictor of students' motivation. A positive affective climate that promoted risk-taking was positively associated with students' mastery orientation, help seeking, and positive emotions associated with learning fractions" (483). This result brings to the forefront what we know to be true: Teachers are powerful agents in affecting students' feelings about mathematics.

Kim's own mathematics learning and teaching experience exemplify Williams's (1980) paraphrase of a Chinese proverb: "Tell me mathematics and I may remember; involve me...and I will understand mathematics. If I understand mathematics, I will be less likely to have math anxiety. And if I become a teacher of mathematics, I can thus begin a cycle that will produce less math anxious students for the generations to come" (101). It is clear that Kim has begun and continues to promote such a cycle.

CONTACT

Kim McReynolds
Cresthill Middle School
9195 S. Cresthill Lane
Highlands Ranch, CO 80126
E-mail: booksandpooh@hotmail.com

UNIT OVERVIEW:
THE GEOMETRY OF KITES

Aim: How can geometric shapes help us to make cool-looking kites?

Objectives: To apply and reinforce students' understanding of basic geometric terms.

Grade Levels: 6–8

Source: Original

Number of 50-minute periods: 6

Mathematics Principles and Standards Assessed:

- Mathematics as equity, curriculum, teaching, learning, and assessment
- Number and operations
- Measurement
- Geometry and spatialsense

Prerequisites: Students can draw and identify angles, parallel and perpendicular lines, polygons and circles.

Mathematical Concepts: Students define and apply concepts of parallel lines, quadrilaterals, perpendicular lines, angles, and circles.

Materials and Tools:

- An overhead copy of the inference "Kite Making"
- Geometric Kites guidelines (Figure 6.2, p. 79) for each student
- Materials:
  - Rulers, compasses, yard stick
  - 2 dowels per student
  - Drawing paper for kite's draft
  - Butcher paper for final kite
  - Construction paper
  - 4–6 different colors of crepe paper for kites' tails

Management Procedures:

- Time Line:
  - Explain project (15 min.)
  - Discuss inference (15 min.)

- Notebook draft
- Draft kite (1–2 periods)
- Final kite (2–3 days)
♦ Procedures:
  - Begin with the inference on making kites.
  - Have students share their guesses in groups.
  - Elicit the aim of the lesson from the students.
  - Make a KWL chart to keep throughout the unit.
  - Distribute kite guidelines to students (Figure 6.2, p. 79).
  - Have students complete the self-assessment rubric (Figure 6.3, p. 82).
  - Have students share the details of their kites.

Assessment: Informal observation, scoring rubric, and journals.

7

MERRIE SCHROEDER: SNACK FOOD CONSUMPTION

I created the Snack Food Unit partly to satisfy one of the missions of my school, which is the generation of new teaching materials and strategies to help colleagues implement Standards-based teaching in K-12 classrooms. The unit is for middle grade students and centers around solving a problem comparing our own eating habits of snack food with other countries. I try to find ways to teach students to develop their own thoughts and ideas because the mathematics classroom is a room full of people, not machines. And people have emotions, and people learn differently, and they feel differently on different days, and most important—they THINK. And they LOVE to think. They love to feel important. This kind of teaching gives students a chance to think and to feel important.

Merrie Schroeder,
Price Laboratory, University
of Northern Iowa, IA

Merrie Schroeder uses numerous resources to create lessons based on real-world data for her classes. She created the Snack Food Unit from statistics reported in magazines published by Snack Food Association, or by Shapiro (1992, 31).

ENGAGING STUDENTS

To begin the unit, Merrie describes and clarifies the goals and activities to come, noting that the first activity—learning all the concepts and gaining the necessary tools—will be done as a whole class, and the application of the concepts and skills will be done in teams. To introduce the first part, she presents interesting questions for students to pursue about snacks: "I have found some statistics about the amount of snack food that people eat in a year in several other countries. I was curious...how do WE compare as seventh graders to these statistics? What are we going to have to do to compare ourselves?" Some students quickly respond that students should start recording how much they eat. In a short time, someone asks, "What IS snack food? Does an orange count? Or do we keep it to chips and candy and stuff like that...stuff our moms don't want us to eat?" Another student counters, "I snack only on healthy foods. I like those banana chips and the dried stuff instead of potato chips. I think my healthy snacks should count." "Yes," another student says, "but most of us don't eat that stuff. Ms. Schroeder, did they say what they counted as snack food?" Merrie answers by quoting from the book: "From potato chips to pretzels, popcorn to pork rinds, the United States leads in snack food consumption among all countries for which data is available. What can you infer from that?" A student says, "Sounds like they're talking about the junk food line." This discussion ensues for a few more minutes until a call for a vote on the definition is made. Students agree that they are going to define snack food as items like chips, candy bars, and the like, and not so much like the healthy foods. The next problem is figuring out how to record it. Given that the book's data is given as the number of pounds of snack food consumed per capita annually, Merrie asks the class to determine how they will make comparisons. Students' comments include:

Marvin: We could weigh ours each time and add it all up.

Mimi: We could use the information on the bags and jars...like if the bag says it holds a pound and I eat half of it, then I record a half pound.

Justin: Can we include our family? They eat a lot!

Merrie: No. let's keep it for our class. Besides, we want our numbers to be smaller, don't we? Won't that make us look like we're healthier?

Joseph: Are we going to do this for a whole year?

Tiana: I think that, if their data is for a whole year, we have to do something to compare it—right...I know! Let's divide theirs by 12 and it will be for a month. Then we can do this for a month.

Joseph: Or we could find out how much we ate in a day and multiply it by 365 to be a year.

Germain: Yeah, but what if you have an off day and really pig out because you got a bad grade on a test or something and you know your parents are really gonna be mad?

Julian: Well, somebody else might not eat anything because they are sick, so won't those make up for each other? We're comparing as a class, aren't we?

After a few more minutes of debate and discussion, the class decides to estimate their snack food eating habits for a year, since that is what they would compare against. Merrie assigns homework requiring students to record what they normally eat for snacks in a day and to check package labels and do whatever weighing is necessary to determine an estimate of their daily consumption. Then they are to compute the estimated yearly consumption. Before they leave, she tells students to make an estimate in their journals of their yearly average. She informs the students that the next couple of days will be devoted to better understanding how to best organize and present the data they will produce. She describes and clarifies the goals and activities to come, noting that the first activity—learning all the concepts and gaining necessary tools—will be done as a whole class, and the application of the concepts and skills will be done in teams. She distributes a handout outlining the goals and procedure of the lessons to follow (see Figure 7.1).

The next day, students come prepared with estimates and are eager to compare what they consume with others in the class. Before class officially starts, students are already drawing conclusions about why their numbers are so different from others. The data's wide range, 40 to 370 pounds, stems from body size to inaccurate estimates. Merrie asks students to compare their numbers to their estimations in their journal. "How close are you?" she asks. Fun discussions fill the airwaves. She asks students to think of reasons for such a range of estimations. As students report some data, she asks them whether the data is for daily or yearly consumption. While some students are unsure about the process for conversion, others, like, Bernadine, are eager to share thoughts:

I looked on the chips and cookies boxes and they said that a serving was 1 ounce. I figured I ate about 3 servings of that stuff a day, so I took 3 ounces times 365. That was 1095. But when I told my mom what we were doing, she said I'd better go on a diet or divide it by 16. That gave me about 68 pounds.

**FIGURE 7.1 GOALS AND INSTRUCTIONS
FOR DATA GATHERING AND ANALYSIS**

---

1. Be able to understand, use, and make:

   • Tables

   • Graphs: bar and circle (pie charts)

2. Be able to understand and compute averages (measures of central tendency):

   • Mean

   • Median

   • Mode

3. Use estimation skills.

4. Work as a team.

To reach the goals, you will:

1. work as a class to learn about and use tables, graphs, the measures of central tendency, and estimation skills. *Topic:* Snack Food

2. Create a project as part of a team to extend your knowledge and skills. *Topic:* Your Choice

Assessment:

1. Evaluation of the team project.

2. To show how well you have reached the goals, you will be given an inventory to complete on your own.

Project:

1. You and your team will choose a topic for which you will gather data from your classmates.

2. You will organize the data in tables and graphs.

3. You will draw conclusions from the data.

4. You will make a presentation to the class.

5. You will be critiqued (in a very kind way) by your classmates to help make your understanding and skills the very best possible.

6. Each of you will hand in your own tables and graphs, even though you developed them as a team. The quality of what you hand in will depend solely on you.

To be sure that all students understand the various steps, she asks Bernadine to explain why she divided by 16. Merrie then instructs the class on the steps to follow and emphasizes the key concerns on unit measures:

> All of you need to be sure that your answers are in pounds. Before we can compare, our data must all be in the same units. Double-check your work: Did you add up all the same unit measures—ounces to ounces, pounds to pounds, and so on? If you added up all ounces, did you divide by 16? Remember that our goal is to compare our class to the world, rather than *you* personally. Do you think that your own number will end up being the one that represents our class?

Students are quick to make suggestions and eager to see what the "average" will be. Merrie invites suggestions on how to proceed. A student suggests that they call out numbers for Merrie to record on the board. She selects a student to do so on an overhead transparency, and once done, asks him/her to find an average. Students are not comfortable with numbers being randomly recorded, so one comments, "They should be written in order, like from smallest to largest." With lots of help from everyone, the numbers are rewritten, but Merrie is careful not to erase the original list. As they think of data processing, most students begin to remember from earlier grades how to compute an average. "That's an awful lot of digits to add. Is there a faster way to do this?" Merrie asks. Some students decide it is OK to have different groups of people add up different sets of numbers, and then compile them for a grand total. Merrie divides the data amongst different groups and an average for the class is computed. "Could we have done this without reorganizing our list?" she asks. Students do see that the organized list was not necessary. Merrie next asks what she calls the *big* question: "What does this mean? You say our class's average is 59 pounds a year. But Bernadine's is about 68. What does *average* mean?" This is a most difficult concept for middle grade students to understand. It is very easy to compute, but because it is not understood, students have a difficult time recalling from one occasion to another how to do it. As soon as they are reminded, they quickly jump to the mechanics of computing this mysterious number. Reviewing the properties of the three measures of central tendency (average, median, and mode) is Merrie's next focus.

To help students differentiate between the measures, she creates a situation requiring students to compare and contrast the information given by each measure. She tells them:

> If three statisticians from *Snack Food* magazine came in here and looked at our numbers on the board, each of them might declare a different "average," and they would all be correct. One might say the average is this one (circling the one in the middle of the organized list),

the second person might say it is this one (circling the most often recorded number/s), and the last person might agree that it is the one
we came up with. How do you think they each came up with their answers?

It is here that great fun with mathematics ensues. Because students are great
at guessing how people arrived at their conclusions and love to try to think the
way others do, they enjoy imagining reasonable responses. After a little time is
spent on students deducing the methods, Merrie guides them to the definitions
and meanings of the three measures of central tendency. Returning to the overhead transparency of the original and the organized lists of yearly averages,
Merrie asks a series of questions to help students distinguish between the measures: "How important is it to have our numbers organized from smallest to
largest to determine the mean? The median? The mode?" "How hard would it
be to determine the median from our first list?" Students are able to state that
the order of the numbers is unimportant to determine the mean; to determine
mode, the order is not critical but it helps; to determine median, the numbers
must be in order. However, Merrie comments, "With all this discussion, it is still
possible for students *not* to understand the meaning of the mean average. Students must be able to verbalize that it is the equal sharing of amounts. We call it
the Robin Hood theory: we take from the rich and give to the poor, so that everyone has the same amount. (Students call this 'mean' to the rich.) How many
people are treated meanly? Half? Less than half? More than half? Why does it
vary? Why isn't it always half?" These kinds of discussions dominate the numerical portion of the lesson. Students are then assigned several smaller sets of
data to practice calculating and describing the mean, median, mode, and range.

Focusing on students' attention once more to the variability in the data,
Merrie next plans to highlight other factors that may contribute to the differences. She asks the class to consider if differences might exist between the measures of central tendency for boys and girls. "How can we answer that question?" she asks. Students are quick to suggest dividing the data in two columns
for listing boys' and girls' data, and then computing the three averages.

Having viewed the data from a numerical perspective, Merrie's next goal is
move students toward presenting a graphical approach. She asks, "Some of us
are very visual interpreters. How can we make the data more visual for us all to
understand?" Some students suggest making a bar graph, but aren't terribly
sure how to start. Merrie tells them to cluster the averages, for example, 0–20,
21–40, and 41–60, and determine the number of clusters necessary. Students begin to play with the numbers and discover that there are lots of clusters. They
suggest making the clusters bigger, for example, 0–50 and 51–100, and enter the
into table the number of boys and number of girls having data in each cluster.
Because most students have had experiences with creating bar graphs in earlier

grades, very little review is necessary to get them to do so. Merrie invites discussions to interpret the data and then asks students to think of other categories that would be suitable for comparing to the data. She comments, "My next steps depend on what they choose to do." Because students suggest height, shoe size, age, involvement in extracurricular activities, and hair color, she gives students a recipe card and tells them to record each of these attributes and to return the card to her. These cards will be used for the second part of the unit: the team project.

Analysis of the data on snack food eaten in other countries is the next lesson. Her major goal is to help students understand how to create and interpret pie charts or bar graphs so that students will be able to extend the interpretations to the class's data. It is now that she gives students a handout with data of snack consumption from other countries (Figure 7.2). Data is omitted on five countries that will serve as the comparison group, and a blank pie chart representing their amounts is given. Merrie tells students that the total amount of food for all five countries is about 40 pounds. In the blank near each country, she instructs students to write an estimate and come to a team decision on the amounts and placements on the pie chart. She asks students first to estimate the amounts for each country and then to guess which country a given wedge represents. (Note that the numbers in parentheses are the actual data and are not shown to students at this time.) Each team chooses a recorder to explain the team's decision to the class. After all groups have had shared results, Merrie shows the actual data. She asks questions about the data to reinforce students' understanding of decimals: "Notice that these data are reported to the nearest tenth, but our own class data was not. What suggestions do you have for making the data look similar? How would you round the data to be usable?" Depending on the class's response, she reviews decimal computations as well as when and how to estimate decimals.

To help the class connect circle graphs to bar graphs, she distributes graph paper and has students create a bar graph representing the same data. Students individually think of a suitable scale for the vertical axis and then plot the graph. They return to their teams to develop a single team graph, and Merrie gives each group an overhead pen and transparency to draw and present the graph. Once a group has decides which scale is best, the rest is easy; after each group has presented its graph to the class, Merrie moves on to help students see how a bar graph can be converted to a circle graph. She gives students scissors and instructs them to cut their bar graph, connect the pieces and shape them into a circle with tape. She then shows the original pie chart and the circular bar graph on the same transparency, asks the class to look for connections between the two graphs. "There are no wedges on the bar graph's circle," she says, "Can you suggest a quick way to draw them?" Some students will suggest estimating the point of the center and drawing wedges to each end of the taped graphs.

FIGURE 7.2 SNACK FOOD GRAPH UNIT

This unit is based on the statistics reported in *We're Number One* by Andrew L. Shapiro, as reported in various issues of "Snack World" magazine, Snack Food Association, Alexandria, VA. Data for five of the countries we will study are not given.

Pounds of snack food consumed per person per year, 1987–1991:

Australia	4.4	Denmark	3.5
Finland	___	Germany	___
Italy	17.6	Japan	3.7
Netherlands	2.5	Norway	8.0
Spain	2.9	Switzerland	2.7
Sweden	___	United States	___
United Kingdom	___		

The five countries listed below will be the focus of our study. Given that the total amount of food for all 5 is about 40 pounds per person per year, estimate the snack consumption for each and enter result on the blank line.

A. United States ___ (19.2)   B. United Kingdom ___ (10.0)   C. Sweden ___ (4.4)

D. Finland ___ (2.2)   E. Germany ___ (6.1)

In the pie chart below, the estimated size of each wedge represents one of the five countries. Guess which of the five countries is represented by each wedge. Use a letter from A to E to represent a country.

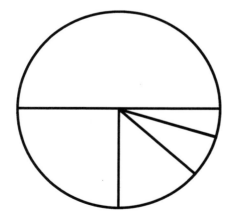

She has the class verify that both pie graphs are now the same and then asks, "Is it possible that different-sized bar graphs for this data will result in the same circle graph? How can we test this idea?" Students suggest checking each other's graphs, and Merrie tells them to do that and to reach a conclusion.

How bar graphs compare and contrast to pie charts is Merrie's next discussion point. At this time, she asks the class for their ideas and lists them in two columns. Students say that they are alike because:

◆ Bar and circle can both be used as percent of a whole.

◆ You can tell the same information in two ways.

◆ Both are graphs.

◆ Both use physical size to determine the largest number.

◆ Both use symbols to identify which amount is how much.

◆ You can use both to compare countries.

Students say they are different because:

◆ One looks like a pie, and one looks like a rectangle.

◆ Bar graphs have a scale of measurement, but circles don't.

◆ Bar graphs have x and y axes, but circles don't.

◆ Circles are good for finding out how much more you need to fill up to a whole (100%).

◆ Circle is easy to do percent for each wedge…you can estimate the percent easier than with bar.

At this time Merrie has students practice changing bar graphs to pie charts. Students are put in teams of four to select any four countries, record the countries' data in a table, transfer it to a bar graph, and then transfer that to a circle graph.

The second part of the unit is to be done in teams. Merrie instructs students to choose their teams and select criteria from the recipe cards to make graphs of their own. For example, a group chooses to analyze the class snack food data by hair color: How much snack food did each blonde eat? Each brunette? Each redhead? Another group looks at snack food consumption by range of shoe size. Students are to develop tables, bar graphs, and circle graphs and use sound mathematical language when discussing the project before the whole class. They should plan to tell not only what the three measures of central tendency are, but to describe the meaning of each and what the implications are. In essence, students are to *tell a story* about their findings, not simply report them. The class has the job of helping students by telling the presenters what they like about their work and asking important questions from which everyone can learn.

Merrie distributes a handout summarizing the team project and assessment criteria (see Figure 7.3). She emphasizes that all presentations should clearly demonstrate the items listed. She adds the discussion requirement because she believes that while the presentations help students prepare for the final exam, the discussions of the presentations allow students a chance to polish their skills.

A written unit exam consisting of five to six questions, which is to be completed by individual students, is the final activity. See Figure 7.4 (p. 100) for a sample exam with a student's answer given in parentheses. Because some of the questions may depend on an answer from a previous question, Merrie allows partial credit to reward valid thinking that may be based on a false premise. For example, Problem 1 presents data with a circle graph of five wedges, each representing a type of fruit. The amount of fruit in one wedge is given, and students are to deduce the number in the others and make a bar graph based on their estimations. If a student incorrectly estimates the sizes of the wedges but has a bar graph that corresponds to the incorrect wedges, then the student loses credit for the wedge part but receives credit for the bar graph.

The examination covers the entire range of Bloom's taxonomy of levels of thinking: recall, application of knowledge, analysis, synthesis, and evaluation.

DISCUSSION BETWEEN COLLEAGUES

What are reactions of the students to your unit?

> They always love this kind of teaching and learning. My students know that they are teachers, valued members of the group, and that I, too, learn from them. They don't hesitate to offer a new idea, ask a different question than was expected, or try to derail the class. Many are still pubescence, too, and love to play and joke and horse around. That is part of growing up. We take that in stride, give the appropriate chagrined looks, regain some control, and move on. Some days are smoother than others, if for no other reason than a change in barometer. Sometimes it's more serious. The math classroom is a room full of *people*, not computers. And people have emotions, and people learn differently, and they feel differently on different days, and most important—they *think*. And they *love* to think. They love to feel important. That is what this kind of teaching does. We have fun while we learn. I try to create opportunities for students to "show off" their knowledge. I want my students to realize that the side trips we take to understand a topic are as valuable as the main trip itself.

FIGURE 7.3 TEAM ASSESSMENT INSTRUCTION
---

1. **Planning:** You will use the information from our class to make a presentation about snack food consumption. As a team, discuss which category you wish to present and how sophisticated you want your project to be. Decide on the number of categories, how you will report your statistics and present your narrative, and so forth. You will have two to three days to complete this part of your work, and one more day to add on any extras that will enhance your project. Although you will work as a team, you will each produce your own copy of the product for your folder.

    When all projects are completed, your team will present its data to the whole class. Class members will be given an opportunity to respond to your work and to ask questions of your team. Be prepared to be in charge! You will hand in your work as a team. It will be scored individually, and will be added to your work folder.

2. **Project delivery:** Hand in the project by teams, with each person handing in their own data. Include one copy of the team's raw data.

3. **Teamwork assessment:** "How did your team do? How did you do?" Include this reflection with the project in the folder.

4. **Presentations:** Each team will tape its graph on the board and conduct the discussion. Be prepared to respond to questions, comments or ideas from the class.

5. **Assessment will be based on:**

    A. Degree of accuracy in presenting what was intended

    B. Degree to which circle and bar graph data match

    C. Completeness of supporting information

    -   Graph title
    -   Labels on bar graph axes
    -   Labels on data in circle
    -   Description of what you try to show
    -   Where the data comes from
    -   Measures of central tendency
    -   Neatness and appeal

FIGURE 7.4 STUDENT'S WORK ON INDIVIDUAL ASSESSMENT

The following circle graph shows how many liked the named fruit best, out of 120 people. Given that 40 people liked apples, estimate the number of people for other fruits and make a bar graph to match the information on the circle graph. Label the axes.

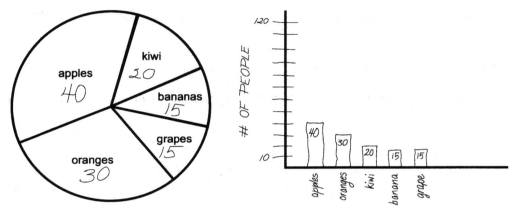

1. How many degrees would be in the wedge labeled "grapes"?

    45°

2. Write a calculator sentence that would give the number of degrees in the wedge labeled "kiwi."

    20 × 360 ÷ 120 =

3. "The average American loves kiwi." This statement is statistically true. Explain.

    The average of the pie graph is 24, the closest number to that graph is 20, which is about how many people said they liked kiwi.

4. Name three ways circle graphs and bar graphs are alike.

    Both are graphs; they can be used to find percents; they use size to show the largest number.

5. Name three ways circle and bar graphs differ.

    Circle graph is a circle and bar graph has axes. To add more groups to a graph without making each group too small, width of bar graph increase but area of circle stays the same.

Question 3 in the individual assessment requires the calculator. How was it used in the project?

> They were not only used, but also studied extensively enough to promote the notion that there are many forms of number sentences that allow us to reach an answer, and that students must have some knowledge of order of operation, number sense, and how programmers design calculators to compute. I also use it to connect wedges by finding fractional parts of the circle. We do stop to do the explorations needed to be able to write the calculator codes that will allow students to find the degrees for the wedges. We use the Explorer calculators, and we really *do* explore. For brevity of this lesson, the actual transition to the mathematics of the circle graph, using calculators, fractions, and 360 degrees, was eliminated. But the same types of investigations and inquiry continue to bring students to calculating the actual degree sizes of each wedge for the countries they choose to examine.

How do you weigh the various responses on the exam?

> Responses are scored in a variety of ways: The short-answer-completion questions net a point each; the essay-style questions are scored at one point for each correct and unique answer. Thus, the exam has no anticipated top score prior to administration. I anticipate the quality of answers and corresponding point value prior to the exam and expect that the average understanding of the concepts and skills would net a student around 8–12 points. Students who have a high-level command of the concepts might score about 17 points. The mark of a student in trouble would be one who does not achieve as many as six points. As it turns out, the class scores are generally good: The scores range from 9 to 17, with a mean of 15.9, a median of 14, and modes of 12, 13, and 14.

How often do you do this kind of project with class? Do you ever lecture?

> I do this type of lesson most often. Lecturing is woven into the lesson when students cannot possibly discover something that is contrived or defined by mathematicians. I think one of the errors some make in mathematics reform (or any other) is to make blanket rules, like "kids should discover meaning." Sometimes kids can't possibly discover meaning because the guy who invented the rule or algorithm or relationship spent his/her whole life doing it in the first place. Then we think we can have a 12-year-old do it, too, because someone did it. Kids can make connections, given the right set of ideas to examine.

And kids can look for patterns, and they can predict and conjecture and test their ideas. But there are lots of places where teachers have to do some lecturing to bring traditional knowledge to the students. Then it is the teacher's job to craft the use of the knowledge so that students can latch it to what they know and make new mental structures to accommodate the new information. *That* is the fun of teaching. And then there is the reality that sometimes just doing plain old arithmetic computation is flat-out fun! Students like a change of pace, and to be able to sit down and just compute is fun. In a way, it is like being a musician…it is greatly rewarding to be a part of a band, adding to the whole piece of music. And other times, it is a real kick just to go off on your own and play a bunch of notes however you wish, just for the joy of doing it. I think math is the same way.

But Merrie, don't you have external constraints that dictate what you must cover and when? What about standardized tests?

I have a curriculum guide, but I do not have to follow a specific time line. Our mission as lab teachers, by Iowa law, is to generate teaching units for Iowa educators. I *have* to do this. We do administer standardized tests. Since those tests do *not* measure programs, we do not use them this way. We use them to look at how students compare at their grade with other students at their grade.

Do you give traditional exams?

I'm not sure any more what a traditional exam is. Therefore, I guess the answer is no. However, I do have pieces of exams that have some tradition in them, such as a section where students just compute…naked math, as it were. When I assess traditionally, it is in down-and-dirty, quick-check quizzes to give me some needed information prior to teaching a skill or concept. At one point I thought it would be important to give students traditional exams so that they could practice this type of activity for success in later courses. But it is such a waste because the traditional (meaning matching or fill-in-the-blank or quick computation) exams don't give *me* any information worth spending the time and energy on. They certainly are easy to grade, but it's about as meaningful as counting all the food in my cupboards to determine how much I weigh. And my students love the other types of exams so much that they don't like the traditional style.

Do you assess your teaching?

I assess my teaching through reflecting on the lesson after I have done it, as well as through revisiting the tests and projects. Another factor is conversations with colleagues. It is important to have opportunities to share with others and to have them provide feedback and ask questions to take the quality of teaching a step further. My colleagues always ask good questions. Those people are hard to find! Having a colleague who looks at my teacher's materials, gives it a cursory glance, and grunts, "Cool," is not helpful. I am fortunate to have colleagues willing to devote the extra time to help me improve my teaching. Another way to keep assessing my teaching is by getting on programs for conferences with a colleague who wants to share my techniques. Preparing for public scrutiny with a colleague *really* makes you polish your work and think hard about why you do what you do. It's different than doing a "show" by yourself. You can bluff your way through a presentation when you are by yourself, but sharing the teaching with a colleague makes you try harder to look good for both of you, and so you find yourself assessing yourself more. Case in point: Having to write the unit for this book really made me do some reassessing. *Snack Food* wasn't very well written, especially for publication. Most of my writing is just to give myself reminders of what I do. When it becomes public, I have to get critical of what I do.

Comment on the units you teach in class. Any reactions from parents?

I have developed all my units. I cannot teach primarily from a text. However, I do use texts for resources of practice problems. I share my units through conferences, and some are earmarked to be written for publication. Our feedback from parents has always been very positive and supportive.

Tell us a bit about your school's minority participation in your classes. Do you notice any differences in their performances on your exam?

Our school has minority students, and they participate as successfully, if not more so, than the others.

Any advice for readers trying your project for the first time?

Advice for beginning teachers: know your mathematics. Many times beginning teachers are learning the mathematics along with their students from very good texts. This is an important learning period in the early teaching years. "Teaching on the fly" comes when there is a great command of the content as well as great comfort in management. Students get excited about topics that relate to them, and their excitement shows physically. A teacher has to be very relaxed about both behavior and ideas that are unexpected. This project may not be written carefully enough for a beginning teacher to venture into. A beginning teacher may need the help of an experienced teacher.

COMMENTARY

Part of Merrie's philosophy of teaching is clearly revealed through a comment she makes while teaching this unit. Very early in her unit, after she asks the class to suggest ways for organizing their own class's data set, she says, "My next steps depend on what they choose to do." While this constructivist response shows Merrie's confidence in her mathematical knowledge, it also reflects her belief that it is her responsibility to understand and appreciate the needs and capacities of her students.

Nussbaum and Novak (1982) argue for teachers to create situations where students are required to invoke, describe, and debate their conceptual framework, and then to support the most generalizable solution. To some extent, Merrie's students engaged in all of these processes when they made estimates and presentations and then had to agree on a best model for the graphs. In effect, they participated not only in mathematical debate but also in scientific inquiry.

The project is one that demonstrates students' strengths, weaknesses, and ability to make connections between different modes of representing data. Building on that information, Merrie guides teams of students to make their own representations of data and present them to the class. She fully integrates problem solving, reasoning, communication, and connections in mathematics as the vehicles for reaching the goals of her unit. But how sound is the unit's mathematical content? A little scrutiny quickly reveals that its mathematical content is very rich. Merrie craftily gets students to unravel the concepts and connections of pie charts and bar graphs while they review computations for averages, uses of ratios, and the rounding of decimals to make better sense of data. Noteworthy is the fact that she does not present the practice of traditional basic skills as "naked math," or isolated skills for students to memorize. Rather, they arise as skills that need to be mastered because they are useful for completing a problem of interest to students. In short, her instructional approach reflects NCTM's view of problem solving as the heart of mathematics: The mathematics that her students do comes from a need to resolve a problem. She mentions that she is free from the pressures of standardized exams. Her unit, however, is ideally suited for instruction under any testing situation. Hopefully, many more teachers will have the freedom to focus on student learning as the NCTM's *Standards* become more widely implemented at all levels. Finally, Merrie comments on how collaboration with colleagues is helpful for generating and polishing her ideas. She supports the importance of collaboration for implementing reform efforts.

Contact

Merrie Schroeder
Price Laboratory School
University of Northern Iowa
Cedar Falls, IA
Phone: (319) 273-5909
E-mail: merrie.schroeder@uni.edu

UNIT OVERVIEW:
DATA GATHERING AND ANALYSIS:
SNACK FOOD UNIT

Aim: How does our class's consumption of snack food compare to other countries?

Objectives: Using data from the Snack Food Association, students will gather, examine, and apply their own data to pie graphs and bar graphs.

Grade Levels: 6–8

Mathematics Principles and Standards Assessed:

- Principles for equity, curriculum, teaching, learning, assessment, and technology

- Mathematics as problem solving, communication, connection, reasoning, and representation

- Number and operations

- Measurement

- Data analysis and statistics

Prerequisites:

- Ratio and proportions

- Degree measures in a circle and related sectors

Mathematical Concepts: Students will apply concepts of mean, median, mode, to create graphs for real-life data. They will determine relationships between pie charts and bar graphs and apply them to the conversion of one graph into the other.

Materials and Tools:

- One copy of Snack Food Graph Unit sheet per student (Figure 7.1, p. 92)

- Masters for transparencies

- Recipe cards for recording information about self and estimated snack foods

- Graph paper, scissors, tape

- Calculator

Management Procedures:

♦ Major sections of unit:

- A. Using statistics on world consumption of snack food (*Snack World* magazine, various issues, published by Snack Food Association, Alexandria, VA, or *We're Number One*, Andrew L. Shapiro, New York: Vintage Books, 1992, 31), students will gather their own personal data on snack food consumption as a class and compare it to other countries. To determine how different their class is, students will learn to record data in tables, use measures of central tendency (mean, median, mode, range) to discuss the data, and represent the data visually in graphs (pie and bar).

- B. Students will work in teams to create a statistical project to show to the class, based on data collected from classmates. Topics will be chosen by students and presented to demonstrate their knowledge of gathering, recording, and analyzing data. Classmates will help critique the presentations.

- C. Students will take a final examination by paper and pencil, individually. The final examination will reflect the knowledge, skills, and concepts gained through studying as a class about snack food consumption and creating team projects.

Assessment: See B and C under Management Procedures.

8

KIM LEBLANC AND DARLENE MORRIS: DISCOVERING BASIC PROPERTIES OF GEOMETRIC FIGURES

We began working together when the Louisiana Systemic Initiative Program funded a reformed-based project in our area. We live in two small towns that are approximately 20 miles from each other. However, we regularly communicate through e-mail, fax, and phone. We talk weekly about the lessons we are working on and the problems we are having with the lessons in our classrooms. Such collaboration has helped both of us implement each other's ideas and suggestions to become stronger teachers in developing lessons to share with others.

Kim Leblanc,
J.I. Watson Middle School,
Iowa, LA

Darlene Morris,
Maplewood Middle School,
Sulphur, LA

Through Kim and Darlene's efforts to implement the NCTM *Standards*, they found the constructivist approach taken by the *Visual Mathematics Curriculum* (Foreman and Bennett, 1996) a great tool to help them promote goals of the *Standards*. One of their favorite units from the text is one where students explore conjectures and form generalizations about angles and triangles.

ENGAGING STUDENTS

They begin by challenging students to form and compute the area of as many triangles as they can with a base of four on a 25-pin geoboard. Students count the geoboards square units, approximate some areas and keep track of their data to discover that triangles with areas of 2, 4, 6, or 8 can be formed. However, some students collaborate and place two geoboards together to find additional possibilities. Directing the students' attention to the base and height, the teachers ask students to try and connect the measures to areas of other figures they previously studied. The idea that the base and height form a rectangle that doubles the triangle's area becomes readily apparent as students examine the data. Not having any formalized rule, students proceed to test this conjecture to in other examples.

The teachers' next challenge for students is to use geoboard dot paper to extend their investigation to triangles not restricted to the 25-pin boundary. Students continue with the base of four units until they reach consensus about the relationship among the base, height, and area of a triangle. Rather than compute square units on the geoboard dot paper, many begin to use the rule that the area of a triangle is half the product of the base and height. An exciting exploration for students is to show that a very long and "skinny" triangle with a base of 4 units and height of 1 has the same area as a "fat " triangle whose base and height are each 2 units.

To further strengthen student's understanding of areas of triangles by having students apply the ideas in different ways, the teachers give students triangles with *no* measurements, and invite them to request minimum information needed to find the area. The teachers write, "We allow time for students to develop their own observations and generalizations from these figures. Together as a class, students come to a consensus on acceptable meanings. We feel this approach to learning is more beneficial to students' conceptual understanding than merely giving them a sheet with the information. The next topic the teachers present relates angles to their measurements. Through group and class discussions, students compare meanings of acute, right, reflex, straight, vertical, and adjacent angles, and then fold circles to form protractors. Because the right angle was previously used in triangle explorations, the teachers instruct students to fold a paper circle in fourths as a first step in creating a paper protractor. Students use their paper protractors to draw a 45° angle, which helps attach meaning to protractor and angle measures. As students work, the teachers observe and are able to see which students are developing some understanding of what is meant by angles, and whether a sense of angle measurement, as a measure of rotation rather than distance, is evolving. Students next use standard protractors and straightedges to draw and measure various angles.

To further assess students' understanding, the teachers ask them to use a standard protractor and straightedge to make drawings that may or may not satisfy given conditions. Some examples students try include:

1. Construct a polygon with the fewest number of sides possible and with at least one angle of each of the listed measures.

   (a) 69°, 10°, 11°

   (b) 90°, 110°, 120°

   (c) 150°, 110°, 90°, 130°

2. Construct, if possible, one or more polygons that satisfy each set of conditions. If a polygon cannot be constructed, Explain why: (a) This triangle has 2 right angles. (b) This triangle has exactly 2 lines of symmetry. (c) This quadrilateral is concave. (d) This triangle has sides of lengths 2, 3, and 7 linear units.

3. For each of the given conditions below, use a standard protractor and straightedge to make drawings that satisfy the conditions, if possible. If not possible, explain why: (a) 2 acute angles that are supplementary, or (b) An obtuse angle and an acute angle such that the measure of the obtuse angle is three times the measure of the acute angle.

Students label the measures of each angle drawn and show or explain how their drawing satisfies the conditions. Decisions on whether students are ready to begin the next unit on exploration of symmetry resides in the explanations of the problem students choose to prove or disprove, and on the accuracy of the measures they find.

DISCUSSION BETWEEN COLLEAGUES

Please comment further on how you generally assess students and when you use traditional assessment.

The idea of interviewing students within the class period and documenting growth is part of what we use in daily assessment. When assessing students, we want to see them making extensions, asking further questions, and computing the calculations correctly. We try to keep accurate records on a clipboard that help us keep aware of who is asking thought-provoking questions, who is sharing their thinking at the overhead, who is building a model to help their group understand a certain problem, and who is ready for an extension. As we strive to move to a more holistic approach to assessment, we will better be able to assess where each student is along a continuum.

In general, how do you perceive your roles in the class?

We see teachers as facilitators and therefore, we are better able to meet the needs of each child. As facilitators, we create a broad agenda that is very flexible to adapt to student needs. The students' interests and questions are important to the direction of the lesson. We ask open-ended questions in an effort to enable the students to construct understanding of a concept or idea. The students are encouraged to discuss ideas within their groups, and to move from concrete to abstract ideas. We might explore different ideas in different classes or even in different groups within the same class. This allows students to excel during parts of the lesson while having their ideas valued in discussions.

Describe the minority population in your school and their participation in your explorations.

While Watson Middle has a minority population of 18% and services a rural community, Maplewood Middle has a minority population of 10% and services a suburban community. In both schools, the *Visual Mathematics* curriculum is integrated in grades 5–8.

COMMENTARY

Kim and Darlene's approach to this unit on angle measurement does an excellent job of taking a traditional "measure, then add or subtract" lesson to the heights of nonroutine problem solving. The unit reflects the *Principles and Standards* for geometry and measurement, and adheres closely to research on geometry recommending appropriate ways to help students reach the level of formal proof in geometry. Through research investigating geometric thinking, Fuys, Geddes and Tischler (1988) showed that a model for the learning of geometry developed by Dutch mathematics teachers Pierre and Dina van Hiele largely describes the way students learn geometry. The model identifies four levels of thinking in geometry. At level 0, the *recognition stage*, students can recognize, construct, draw, or copy shapes as a whole. At level 1, the *analysis stage*, they are able to analyze shapes, formulate their definitions, and discover their properties. At level 2, the *ordering stage*, students are able to use the definitions and properties of figures to formulate informal proofs about these figures and their properties. At level 3, the *deduction stage*, students formally prove theorems deductively, and at level 4, the *rigor stage*, they can rigorously compare different axiomatic systems. The model asserts that students must progress through one level before attaining the next, and that progression through the levels is determined not by chronological age, but by appropriate instructional experiences. The study supported the idea that each level has its characteristic language, and that instruction at an inappropriately high level may explain some of the difficulties that secondary school students experience with geometry. Results of the study also demonstrated that the use of concrete materials and manipulatives can motivate students to play with their ideas and also help students become more reflective as they process varying ways of thinking.

The middle grades should provide ample experiences for students to engage in geometric thinking at the first two levels. This unit is a good example of how this can be done: Students use manipulatives that clearly demonstrate and measure the essential properties of angles and area (level 1); they continually search for patterns to formulate and discover formulas or relationships (level 2). The progression of activities and experiences in this unit helps solidify students' understanding of the topic, while preparing them for later success with formal geometric thoughts.

Collaboration of teachers to plan for reform is not easily accomplished even within a single school. The fact that Kim and Darlene continue to collaborate from schools that are 20 miles apart is to be highly commended.

CONTACT

Kim Leblanc
J.I. Watson Middle School
Calcasieu Parish School Board
201 E. First Street
Iowa, LA 70647
E-mail: kim.leblanc@cpsb.org

Darlene Morris
Maplewood Middle School
Calcasieu Parish School Board
4401 Maplewood Drive
Sulphur, LA 70663
E-mail: dmorris@linknet.net

Unit Overview: Discovering Properties of Basic Geometric Figures

Aim: What are some properties of angles and triangles?

Objectives: Students will use a geoboard, protractor, and straightedge to formulate conjectures about angles and triangles.

Grade Level: 6

Source: The Bennett, J., and Foreman, F. (1996). *Visual Mathematics Curriculum*, Mathematics Learning Center, Portland Oregon.

Number of 50-Minute Periods: 3

Mathematics Principles and Standards Assessed:

- Principles for equity, curriculum, teaching, learning, assessment
- Mathematics as problem solving, communication, connection, reasoning, and representation
- Number and operations
- Algebra
- Measurement
- Geometry

Prerequisites:

- Concepts of angles, triangles, polygons, area
- Supplementary and complementary angles

Mathematical Concepts: Using geoboard and geoboard dot paper, students will compute areas of triangle, formulate and test the area formula for triangles, measure and construct supplementary and complementary angles, and construct polygons satisfying given conditions.

Materials and Tools:

- 25-pin geoboards, one for each student
- Geoboard dot paper
- Paper circle
- Standard protractor and ruler

Management Procedures:

- Challenge students to use geoboards to compute the area of as many triangles as they can with a base of four units. Have them formulate a hypothesis about base, height, and area of triangles. Extend to geoboard dot paper and have them test their conjectures.

♦ Have students make and use paper protractors of 90° to approximate measures of various angles.

♦ Assign students problems to apply area formula and degree measures of polygons.

Assessment: Circulate to observe and question students' work. Have students use a standard protractor and straightedge to make drawings, which may or may not satisfy given conditions.

9

JIM SPECHT:
THE LAST GREAT RACE

The Iditarod Trail Sled Dog Race is as much a part of the culture of Alaska as Mardi Gras is of New Orleans. The high school principal in McGrath, one of the race checkpoints, once said that their town only observes three holidays—Christmas, the Fourth of July, and Iditarod. When I learned that the race is covered on the Web in real time for two weeks every March, I thought it would be a great project for my general math students. Caught up in following the race, they could learn how to interpret data in various kinds of tables and graphs, and at the same time we could explore the geography and culture of Alaska.

Jim Specht,
Hillsboro High School,
Hillsboro, OR

Jim Specht tries to use real-world activities as much as possible in his math classes. For his general math students, he finds it's also important for the activity to be intriguing; otherwise, it's just another word problem. Before he assigns a lesson, he tests it on himself. Is it interesting? Will he enjoy seeing the students' work, or will it be just another set of papers to grade? For a long-term project, is the math component real or contrived? The idea of following the Iditarod —the *Last Great Race*—with his class struck him as "just so cool" that he decided to give it a try.

Jim's preparation for the project was extensive. One concern was that his department has only one on-line computer; in addition, he may get a busy signal as he tries to access the Iditarod Committee's homepage (www.iditarod.com) during class. He planned to use his home computer as a backup, downloading the latest data before coming to school each day.

He collected, and placed on reserve in the school library, periodicals with reports on previous races. He purchased from the Iditarod Committee a videotape of the race; from AAA, a map of Alaska; and from an airport, an airline chart showing rivers, elevations, towns, airports, and radar beacons across the state of Alaska. Based on the data available on the homepage, he devised a number of appropriate and interesting questions for his students to investigate. One week before the race, he taught (or reviewed) how to create and interpret bar graphs, line graphs, and box-and-whiskers plots, using a graphing calculator.

ENGAGING STUDENTS

Now he is ready to introduce the project to his students. He starts with a general discussion. "Who has heard of the Iditarod Trail Race?" (Some share bits and pieces of information.) "How did it get started?" (No one knows.) "What's a musher?" (The person who handles the sled and the dogs.) "What must it be like to travel by dog sled for 1,131 miles, from one coast of Alaska to the other?" And, having piqued their interest, "How can we find out more about the Iditarod?"

He tells the class that they will be immersed in the race for the next two weeks. They will download the latest information from the Internet every day, and prepare graphs, tables, charts, and maps from the data. Each will choose two individual mushers to follow, keeping a daily log of their adventures and analyzing the race data to explain their performance. Each student will compile a folder for the project (see Figure 9.1).

FIGURE 9.1 IDITAROD FOLDER REQUIREMENTS

All work completed for the project will be organized in a folder. Check all calculations and hand in neat work. Please organize your work according to the sections below.

1. Title page
2. Table of contents
3. History of the Iditarod
4. Biographies of two mushers
5. Map of Alaska with the trail outlined
6. Graphs of:
    a. Male vs. female competitors
    b. Veterans vs. rookies
    c. Elevation of sites
    d. Daily high and low temperatures in Anchorage
7. Daily log sheets
8. Evaluation of project

The first assignment is to research the background of the race. Students turn first to encyclopedias ("of course," says Jim) and *National Geographic.* "They didn't think of the Internet itself as a resource—for many, this project was their first experience gathering data from the Web." Sharing what they find, they submit their reports. Most have some version of the following story:

> In the winter of 1925, the Alaskan community of Nome was threatened by a diphtheria epidemic. The townspeople feared that, without help, the settlement would be devastated. Pack ice had frozen the harbor, and in an era of biplanes and model-T cars it appeared that there was little hope that help might arrive before the spring thaw. When the people in the town of Anchorage learned of the threat to their fellow Alaskans, heroic mushers volunteered to use dog sleds to take the necessary medicine across the 1100 miles of roadless mountains and frozen tundra. Their hazardous journey through blizzards and ice storms over the Iditarod Trail saved the settlement of Nome. To this day, it is honored yearly by a dog sled race that covers approximately the same route. Many people in Alaska follow the Iditarod closely with great pride, as it is a part of the culture unique to this 49th state.

Next Jim downloads the list of mushers (see Figure 9.2), which gives their names and home states and identifies first-time Iditarod racers. From an overhead transparency, students record facts and statistics about the racers. Jim asks preliminary questions:

- How many of the mushers are Alaskan? What percent of the total is that?

- How many of the racers are Iditarod rookies? Veteran racers?

- How many women are competing this year?

Working in pairs, students examine the data, calculate the answers, and sketch circle graphs by hand to represent the results (see Figure 9.3).

From the table of biographical data, Jim tells each student to choose two mushers. (Some take the time to research past race results in *Sports Illustrated, Newsweek,* and *Time* before choosing "their" racers!) From the data, they write profiles of the two for their Iditarod folders.

The next task is to sketch the race trail on copies of the AAA map of Alaska. Jim downloads the table of checkpoints (see Figure 9.4) and tells students to use that as a guide. But although the map shows Alaska's roads and towns, students soon discover that the Iditarod Trail crosses vast stretches of wilderness, and many of the checkpoints are not actual communities of any size. How can they determine the approximate locations of the checkpoints that are not shown on the map?

FIGURE 9.2 ADAPTED LIST OF MUSHER FACTS

Last Name	First Name	Home Town	Status	Sex	Last Ran	Best Finish	Year
Buser	Martin	Big Lake, AK	Veteran	M	2000	1	1992-4-7
Cotter	Bill	Nenana, AK	Veteran	M	2000	3	1995
Gebhardt	Paul	Kasilof, AK	Veteran	M	2000	2	2000
Halter	Vern	Willow, AK	Veteran	M	2000	3	1999
Hahn	Nils	Germany	Rookie	M	0	0	0
Ramstead	Karen	Perryvale, Canada	Rookie	F	0	0	0

FIGURE 9.3 STUDENT'S CIRCLE GRAPHS

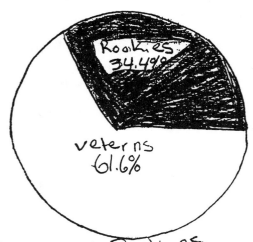

Veterns 61.6%

Rookies 34.4%

Veterns + Rookies Who Raced

Men + Women Who Race

Women 12%

Men 87.9%

FIGURE 9.4 CHECKPOINTS AND DISTANCES (EVEN YEARS)

#	Checkpoints	Distance Between Checkpoints	From Anchorage	To Nome
1	Anchorage to Eagle River	20	20	1131
2	Eagle River to Willow	50	70	1081
3	Willow to Yentna	45	115	1036
4	Yentna to Skwentna	34	149	1002
5	Skwentna to Finger Lake	45	194	957
6	Finger Lake to Rainy Pass	30	224	927
7	Rainy Pass to Rohn	48	272	879
8	Rohn to Nikolai	93	365	786
9	Nikolai to McGrath	48	413	738
10	McGrath to Takotna	23	436	715
11	Takotna to Ophir	38	474	677
12	Ophir to Cripple	60	534	617
13	Cripple to Ruby	112	646	505
14	Ruby to Galena	52	698	453
15	Galena to Nulato	52	750	401
16	Nulato to Kaltag	42	792	359
17	Kaltag to Unalakleet	90	882	269
18	Unalakleet to Shaktoolik	40	922	229
19	Shaktoolik to Koyuk	58	980	171
20	Koyuk to Elim	48	1028	123
21	Elim to Golovin	28	1056	95
22	Golovin to White Mountain	18	1074	77
23	White Mountain to Safety	55	1129	22
24	Safety to Nome	22	1151	0

Jim introduces the idea of interpolating between two known checkpoints to approximate the position. The students have studied ratio and proportion before, but Jim knows they have not mastered these concepts. He guides them through the steps, the answers for which are in parentheses:

1. We need to find approximate locations for White Mountain and Safety. What information from Figure 9.4 and from our map can help us? (The checkpoints are between Golovin and Nome, both of which are on the map. Figure 9.4 shows the distances.)

2. Let's start with White Mountain. How far is White Mountain from Golovin? (18 miles.)

3. The scale of our map is 1 inch for every 60 miles. How can we use that information? (18 miles is about ⅓ of 60 miles, so the distance from White Mountain to Golovin on the map should be about ⅓ of an inch. Draw a 1-inch segment; place compasses [the kind used for geometrical constructions] on the segment and adjust them to a little less than ⅓ inch. On the map, use the compasses to draw a circle with that radius around Golovin.)

4. What other clues can we use to locate White Mountain? (Students find this aspect interesting. The map shows that Golovin is on the seacoast, with mountain ranges to the north. The name "White Mountain" is a hot clue.)

5. What should we do next? (Follow the same procedure for Safety, which is 22 miles from Nome. Use the compasses to approximate a little over ⅓ of an inch, and mark that distance. Any other clues? Nome is also on the coast, just south of the mountain range; the trail should stay out of the ocean!)

6. Finally, check to see if the distance between White Mountain to Safety, as we have located them on the map, is consistent with the information in Figure 9.4.

Having completed their trail maps, students are ready to think about situations that might affect their mushers on the trail. Jim presents this problem on the overhead projector:

> "Suppose a musher makes it from Koyuk to Elim in 6 hours. The same musher goes from Elim to Golovin in 7 hours. Is this surprising?"

Using the information in Figure 9.4, the students set to work. Koyuk is 980 miles from Anchorage; Elim is 1028 miles from Anchorage. The difference is 48 miles; they divide by the time, 6 hours, to find the rate, 8 mph.

Now they look at Elim and Golovin. At this point, one student remembers that the figure shows "miles to next checkpoint," so they can save the step of subtraction. (Jim is pleased; one of his goals was to encourage students to seek patterns that will reduce computation as they read the charts.) The students calculate the musher's rate from Elim to Golovin and find that it's only 4 mph.

A discussion ensues. What might be the reasons for the difference in speed? Students offer ideas: Maybe the same dogs were used in both legs of the race and are getting tired; maybe the musher had to leave some dogs in Elim. What about the geography? Was the team climbing a mountain, or traveling through an area where there is very little snow? What about the weather? They conclude that they need more information. Jim is ready to oblige.

He downloads weather reports. Students make line and bar graphs of daily high and low temperatures in Anchorage and other sites along the trail; this gives them an idea of the trail conditions for the different teams. They use their map of Alaska and the weather reports to draw conclusions about wind direction and clear and cloudy skies.

Another table gives the elevations of the checkpoints which students use to plot the position and elevation of each checkpoint on the airline chart Jim has taped to the wall. On this larger map, the geography of Alaska is vivid—they note the vast mountain ranges, huge tracts of wilderness, great distances between populated areas, and very few roads. No wonder the AAA map doesn't show all the checkpoints! They prepare line graphs showing the elevation changes along the trail (see Figure 9.5), and it becomes obvious why some stretches of the trail go more slowly than others.

The highlight of each day, of course, is checking the racers' daily standings (Figure 9.6, p. 125). About 15 minutes before class, Jim downloads the latest data and hustles to the office to transfer the table to an overhead transparency. Eager to see how their mushers are faring, the students are soon familiar with how to read the table—place standing, bib number, latest checkpoint, with the arrival time, number of dogs, and rest period, followed by the departure time from the previous checkpoint (given in "military time"). Jim may zip out in the middle of class and download even more current data, usually with three or four changes in musher position—information the students pounce on and quickly process. Jim keeps the questions perking: "Is your musher at the halfway point? Past it, or not there yet? A third of the way? Are the dogs making better time today? Is your team gaining on the others?"

FIGURE 9.5 STUDENT'S GRAPH OF ELEVATION POINTS

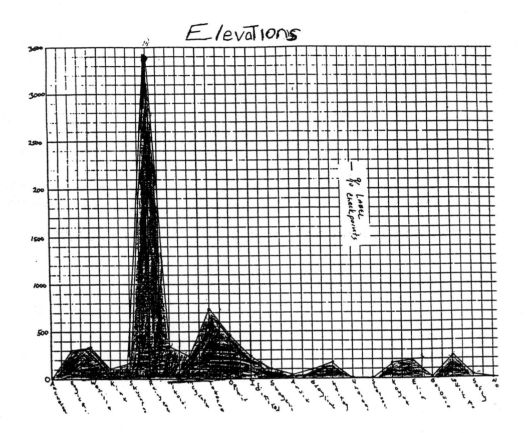

Jim distributes daily log sheets (see Figure 9.7, p. 126), on which students record, compute, and reflect on the information about their two particular mushers. One of the entries is the musher's average speed in the last 24 hours. Students know the formula, but correctly entering the real-time data from Figure 9.6 into their calculators presents a challenge. In Figure 9.7, for example, Vern has traveled 30 miles in 4 hours and 42 minutes. The first step is to estimate; 30/5 = 6 mph. But as the students perform the actual calculations, Jim notices some wildly different results on their calculator screens; in addition to 6.4, he sees 49.5, 65, and 8.2, depending on where students inserted parentheses, or whether they remembered to change minutes into hours.

FIGURE 9.6 ADAPTED LIST OF 63 DAILY CHECKPOINT STANDINGS

#	BIB	MUSHER	ID	Checkpoint Now	IN Date	IN Time	# Dog	Rest Time	ID	Checkpoint Before	OUT Date	OUT Time
1	34	Jeff King	21	Elim	3/16	13:28	8	00:35	20	Koyuk	3/16	7:42
2	3	DeeDee Jonrowe	20	Koyuk	3/16	10:25	9	06:18	19	Shaktoolik	3/15	13:41
3	32	Doug Swingley	20	Koyuk	3/16	3:22	9	06:16	19	Shaktoolik	3/15	19:45
4	37	Martin Bauser	20	Koyuk	3/16	3:17	9	06:22	19	Shaktoolik	3/15	19:48
5	58	Charlie Boulding	20	Koyuk	3/16	4:12	10	5:29	19	Shaktoolik	3/15	19:51
6	4	Mitch Seavey	20	Koyuk	3/16	4:52	11	4:58	19	Shaktoolik	3/15	22:18
7	17	Vern Halter	20	Koyuk	3/16	5:09	11	5:01	19	Shaktoolik	3/15	21:40
8	36	Rick Swenson	20	Koyuk	3/16	5:18	9	5:14	19	Shaktoolik	3/15	21:42
9	18	Linwood Fiedler	20	Koyuk	3/16	6:14	9	4:23	19	Shaktoolik	3/15	22:45
10	64	John Baker	20	Koyuk	3/16	6:11	10	***	19	Shaktoolik	***	20:38
11	49	Ramey Smyth	20	Koyuk	3/16	8:38	8	***	19	Shaktoolik	3/15	23:27
12	27	John Barron	20	Koyuk	3/16	8:38	7	***	19	Shaktoolik	3/16	00:38
13	38	Paul Gebhardt	19	Shaktoolik	3/16	6:16	10	6:07	18	Unalakleet	3/16	00:50

*** Blank line indicates that no reading was registered.

FIGURE 9.7 DAILY LOG SHEET WITH SAMPLE ENTRIES

Daily Log Sheet (with Sample Entries)

Today's Date: _March 16_

Musher #1   Name: _Vern Halter_   Bib# _17_

Position in the field: _7_

Latest milepost: _Koyuk_   Trail Miles: _980_

Last milepost listed: _Shak Toolik_ Trail Miles: _922_

Distance since yesterday: _58 mi_ Time elapsed to travel that distance: _8:31_

Average speed in mph: _6.8 mph_ Calculation: _58mi/(8+31÷60)hr_

Musher #2   Name: _Paul Gebhart_   Bib# _38_

Position in the field: _13_

Latest milepost: _Shak Toolik_   Trail Miles: _922_

Last milepost listed: _unalakleet_ Trail Miles: _882_

Distance since yesterday: _40 mi_ Time elapsed to travel that distance:

Average speed in mph: _5.6 mph_ Calculation: _40mi/(7+6÷60)hr_

Box and Whiskers Plot of Distances (Calculator Printout is ok)

```
        |-----|-----|-----|-----|-----|-----|-----|-----|-----|
       360   435   510   585   660   735   810   885   960   1035
```

Average Distance: _818_   Last Place Distance: _1028_

Median Distance: _792_   First Place Distance: _364_

Anecdotes of Interest: _Paul and Vern are doing better_
than most. The road to Kaltag is easy for many
muskers because most of them get there at
about the same time.

Jim reminds them to check against their estimations—"Does my answer make sense?"—and to check their entries in the calculator. He takes the opportunity to review the order of operations, writing the most common incorrect results on the board and asking students to guess the kinds of errors that produced them.

In another challenging exercise, Jim's students use a graphing calculator to prepare a box-and-whiskers plot of the current standings within a nine-interval range. This requires that they find "friendly" interval lengths before entering the range data. To do this, they subtract 5 from the least number of miles, then try interval lengths in multiples of 25. For example, using unadapted versions of Figures 9.4 (p. 121) and 9.6 (p. 125), students see that Jeff King leads and is at Elim which is 1028 miles from Anchorage, and the last-place team is at mile 364. They calculate: $365-5 = 360$; $50 \times 9 = 450$; $360 + 450 = 810$. They see that 50 is too small an interval to cover the range, so they try 75 or 100. Those using intervals of 75 set the range of their calculators for 360 to 1035.

In List 1 on the TI-82/83 calculator, they enter the current "miles from Anchorage" standings of all the racers, based on the information in Figures 9.4 (p. 121) and 9.6 (p. 125). (Although they could have shortened the process by entering the frequency in List 2 (see Figure 9.8), all quickly enter the data in List 1 without a frequency count.) The calculator generates a box-and-whiskers plot. Looking at the graph, students are surprised to see the box without the typical median line inside (see Figure 9.7). "Why did this happen?" they ask.

FIGURE 9.8 LOADING DAILY STANDINGS

L1	L2	L3
1028	1	
980	11	
922	7	
882	7	
792	24	
698	9	
646	1	
413	2	
365	1	

From Jim's preparatory lessons, they have a general sense of how the lengths of the various sections depict clusters and spread of the data. Now, taking advantage of their curiosity, Jim seizes the opportunity to explore the concept of quartiles. Examining the data in Lists 1 and 2 and the Stats + Calc menu of his overhead calculator, he reviews box-and-whiskers plots, introducing new language. For example, in addition to saying, "Half of the mushers are clustered between these two checkpoints," Jim shows that they can also state, "Between 50 and 75% of the mushers are clustered between these two checkpoints, or quartiles."

As they continue to apply the quartiles, students gradually understand that the "lost median" actually overlaps with the first quartile line, at the left end segment of the box. Jim asks them to interpret this information. Their most common response is that 25% of the racers are at milepost 792, where 24 of the 63 racers have checked in. "Can you justify that from the data?" he asks. Doing the calculations, they find that 25% of 63 is less than 24. Now they realize that they should have said *at least* 25% were at that point. Finally, working together, students copy the graph and report the data from the Stats + Calc menu of the TI-82/83 calculator in their log sheet (see Figure 9.7, p. 126).

Based on the maps, graphs, and charts they are compiling, together with the current weather reports and their daily log sheets, they keep a journal reflecting why they think their mushers are progressing as they do. Some class time is also devoted to storytelling; students share interesting items from their research, and Jim downloads anecdotes published on the Web site each day by reporter Lois Harter, who is following the race. They find their way into the journals, which stimulate class discussions, for example, *SPCA Protests That Race Is Inhumane.* "What do you think?" Jim asks. The class watches a video of last year's race, and sees the harsh conditions under which the mushers and dogs persevere. A lively debate ensues.

By the end of the project, students are dropping in during lunch and their free time to ask Jim to print out the latest update. "Having followed their mushers across the wilderness," he says, "they have a very real sense of ownership." As the race ends, they complete their reports and graphs, and submit the project folder for evaluation. The next day, Jim distributes a form for their assessment of the project. Students are also required to write a personal summary of the project (see Figures 9.9 and 9.10).

FIGURE 9.9 STUDENT ASSESSMENT OF PROJECT

On a scale of 1 through 10 (10 being highest), how would you rate the Iditarod project in the following areas:

1. Maintained your interest. _____

2. Taught you things you did not already know. _____

3. Applied math and graphing skills in a meaningful way. _____

4. Helped you practice organization of materials. _____

5. Was worth the time. _____

6. What part of the project did you like best? _____

7. Offer suggestions for improving the project:

    _____

    _____

    _____

8. Did you choose to share any of this project with your folks? If *yes,* what part? _____

    _____

9. In 3 to 5 sentences, why do you think it is called *The Last Great Race?*

    _____

    _____

    _____

    _____

    _____

10. For homework, write a personal summary of the project describing what you have learned. Would you want to see it someday? Why or why not?

FIGURE 9.10 STUDENT'S REFLECTION ON THE PROJECT

Personal Summary

When we first began this project, I had no clear idea of what the Iditarod Race was, how it got started, or any knowledge of the accompanying details associated with this quest. Three weeks later, I now have a clearer idea of all, and though I still feel that it is taking advantage of dogs in one form or another, I do have a certain amount of respect for the race. It takes a strong person, mentally and physically, with a great trust in just a handful of dogs, to brave the Alaskan climate and terrain in hopes of not only winning the prize money of such a small amount, $50,000, but also being able to have the satisfaction of saying one completed it. Besides the knowledge of the statistical information gained, I found the race more interesting considering that I had two of my own mushers to have a certain amount of personal interest in. Even though neither one even finished in the top 30, the suspensefulness of the race was heightened because I had my mushers to root for.

I had peace of mind to some extent, also, knowing that the dogs were being treated wonderfully. There should not even be consideration of treating them any other way considering the feat they are forced to undertake. I say "forced" because the dogs really don't have any other choice. It has been said that they just love to run and go crazy if they can't, and it could be so, but I wonder who exactly has asked them and if they carried on an in depth conversation with the dogs to obtain their opinion about the Iditarod.

However, overall, the project was less grueling than I thought it would be. Sometimes it was tedious and incredibly frustrating, mainly the first two days, but it got easier with time. I had an enormous amount of help from two incredibly kind fellow students who recorded my information for me when I became very sick. That was probably the single most stressful part of the whole project next to finding the elevations of all the checkpoints.

When the Iditarod comes around next year, I wonder if I'll tune in to it. Either way, my perspective on it will no doubt be different. I could say I might even be interested in it, but the more appropriate word, I believe, would be accepting!

Discussion Between Colleagues

What does a typical day in your class look like?

> We integrate the use of graphing calculators into instruction several times a week. Because our texts are literally years behind the technology, I often have to give oral instructions or handouts to help students work with the calculator. On projects like the Iditarod, I encourage students to rely on each other for help.

Students likely had gaps in their mathematical knowledge. Comment some more on how you prepared them to tackle your project.

> Having them work in teams of two was very beneficial—the gaps were not so great that two of them together couldn't help each other understand the work. I do believe that children are born curious. My first responsibility is to kindle that curiosity into a desire for learning. If they know that I am truly working on it, they will generally try to work with me. When the students are engaged, everything else solves itself—discipline, attendance, attitude, work ethic. What generally erodes students' enthusiasm is the burden of number crunching. I apply a judicious use of technology to keep the momentum going.

How did you assess the folders?

> I graded them one page at a time. It was nothing profound. Every page was worth a maximum of 10 points, and 18 pages were expected, for a total of 180 possible points. I valued:
>
> * Mathematical accuracy.
> * Completeness—Did they finish the page? Include all entries?
> * Neatness and correctness in following directions.

What can you say about what students learned?

> I think they valued what they learned. I require them to include some projects as part of their course portfolio. Every student selects the race for inclusion. When you asked me to send you copies of students' work, I contacted five of my former students and asked them if they still had their folders. All five had not only kept them, but could easily locate them! All students also choose to include this project as an entry for their semester portfolio. Their reflections showed that they learned how mathematics can serve as a vehicle for exploring another place and culture. Finally, when I read reflections that say "the project

was sort of interesting," or "that wasn't bad," I take it as very high praise of a math project from these students.

What about your role as teacher during this project?

> This activity was fun for me. As the race progressed, it was amazing to me to think that up in Alaska one of the judges at a checkpoint would log onto the Internet with a laptop and a cell phone, and 10 minutes later I was able to place a downloaded copy of that data on an overhead transparency in my classroom. We truly live in an age of wonder.

> My role in the classroom also contributed to my enjoyment—not a boring moment for me because students did not all work on the same thing at the same time. This was partly due to students' preferences, and partly due to logistics—for example, not everyone had a graphing calculator. Once students had the current downloaded information, I felt more like a coach. I would do a bit of refereeing to make sure that materials were shared fairly, or point out that several students could work from one box of colored pencils if the colors were shared wisely, and that protractors could be used as straightedges if the line was less than 6 inches long, and so forth.

What did you learn from your students about projects of this sort?

> Unfortunately, it did not prepare them for the climate of some of the traditional courses that came later. When they expect class to be interesting and intriguing, a traditional textbook course doesn't hold much interest.

What factors determine which students are placed in general mathematics?

> The junior high teachers recommend this placement for their students.

Will your general mathematics students be required to take algebra?

> No, but we have three levels of general math. Students are not required to take all three but may move into algebra after experiencing a high degree of success at any of the three levels.

The Iditarod Committee now has all the information necessary for your lesson nicely organized on its home page. They have also provided a map of the trail with *all* of the towns having a milepost in the race. Would you modify your lesson the next time you teach it?

> I would keep the AAA maps. Students learn the geography of one of our states while they apply mathematics to determine the location of unfamiliar places. Both skills are worthwhile.

Do you have any further suggestions for readers?

> Require that students submit photocopies of their resources. I forbid originals because I don't want library materials to suddenly and mysteriously have holes in them. Although I did not know it when I created this project, a variety of classroom packets for this race are already available for use in schools. I also highly recommend that teachers show a videotape of the race—it strongly engages student in the spirit and culture of the experience.

> Finally, dare to be bold. There is no glory in doing something that is easy.

COMMENTARY

Jim's idea to "reveal" (his own word) ratio, proportion, and data analysis to students through the race project did so in a way to deepen students' understanding of the mathematics and its applications. His approach is quite different from the traditional method ("Today we are going to solve proportion problems. Turn to page 101."). The computations and graphs were all integral to the process following "their" mushers. The real data from the race also presented students with a situation that is not often easily encountered otherwise. For example, the box-and-whiskers plot, where the median and first quartiles were the same, showed a noncontrived graph of an event that required students to wrestle with making sense of a definition in light of the data: "What do the median and the first quartile represent in this case?" Although students used a calculator to produce the graph, they needed to determine an appropriate range and connect the table of values to the graph in order to make sense of the graph and its related output.

Jim's approach to proportional reasoning in conjunction with the calculator also required students to do more than merely push keys. Checking the calculators' results with students' estimations revealed how calculators do not free the mind of the thinking processes necessary to tackle nonroutine problems. As an example, to facilitate easy entry into the calculator, Jim chose to show students how to represent rate of speed as: (given miles)/(given number of hours) = x

miles/1 hour. This unit-rate approach for solving rate problems "has intuitive appeal because children have made purchases of one and many things and have had the opportunity to calculate unit prices and other unit rates" (Post, Behr, and Lesh 1986, 8).

Jim's use of the Internet and the newspaper was a nice way of integrating the latest in technology with readily available resources to engage students in a current event outside of their community. Both tools were effective in getting students to see and apply mathematics in making sense of data. Students' sharing of stories about their mushers, or reflecting on the data they collected, provided opportunities for them to engage in communicating and learning about a task that was not only worthwhile but also based in the culture of one of our states.

Soon after he taught this unit, Jim wrote to me:

> I've just returned from speaking at a math conference in Anchorage (+5 degrees and 20-mph wind). I got to meet Martin Busser, three-time winner of the race. He was quite a guy, but the dogs were the real stars because they too were so beautiful and are such superb athletes. Some kids in Alaska collect cards of the sled dogs, much the same way that kids here collect baseball cards, or Canadian children collect hockey cards. In all, I had quite an experience.

Jim's willingness to wrestle with whatever problems came up as he tried this lesson for the first time is to be commended. For other examples of ways to extend learning with technology, see Jim's book on using the graphing calculator—*More than Graphs* (Key Curriculum Press, 1-800-995-MATH).

CONTACT

Jim Specht
Hillsboro High School
3285 SE Rood Bridge Road
Hillsboro, OR 97123
E-mail: spechtj@hsd.k12.or.us

UNIT OVERVIEW:
THE LAST GREAT RACE

Aim: What are factors crucial to winning the Iditarod Trail Sled Dog?

Objectives: Students will collect information on Alaska's Iditarod Race and deduce factors that helped determine outcomes.

Grade Levels: 7–9

Courses: General mathematics, prealgebra

Source: Original

Number of 45-Minute Periods: 9

Mathematics Principles and Standards Assessed:

- Principles for equity, curriculum, teaching, learning, assessment, technology.

- Mathematics as problem solving, communication, reasoning, representation

- Number and operations

- Algebra

- Measurement

- Data analysis and probability

Prerequisites: "Students just need a genuine work ethic. A teacher can attend to skills instructions needed throughout the unit. However, a review of line graph, bar and circle graphs, and box-and-whiskers plots is useful."

Mathematical Concepts: Students use a map's scale factor, distance formula, and ratio and proportions to estimate unknown distances and speed of racers. They plot or interpret circle graphs, line graphs, box-and-whiskers plots.

Materials and Tools:

Note: The race generally begins on the first Saturday in March. Check the Iditarod homepage (www.iditarod.com) for dates. The Iditarod Trail Race Committee also sells classroom kits.

- At least one online computer

- Graphing calculator with list capability similar to TI-82 (for pairs of students)

- Iditarod homepage (www.iditarod.com) to download facts and statistics about the running of the race:

  - List of mushers

- Current daily standings of mushers
- Data on checkpoints and elevations
- Weather conditions
- Newspaper covering the race
- Copy of a map of Alaska for each student
- Airline map of Alaska
- Previous periodicals on the Iditarod
- Overhead projector

Management Procedures:

- Discuss basic information about the race with students. Interesting trivia information and history are available from the homepage.
- Assign students to find additional information to complete a history of the race.
- Give students a copy of the list of mushers and have them graph rookies versus veterans, males versus females.
- Have students:
  - Recreate the trail by using a map of Alaska, a copy of the checkpoint distances from Nome, and compasses.
  - Draw a line and bar graph of low and high temperatures in Anchorage.
  - Draw a line graph of the elevations of the checkpoints.
  - Write biographies of two mushers and track them daily.
  - Complete their daily log sheet.

Assessment: After the race is over, have students submit all entries for a grade and complete an assessment of the project.

10

KATHARINE OWENS, RICHARD SANDERS, AND ROBERT LIPSINKI: GAMES FOR PRACTICE AND REINFORCEMENT OF CONCEPTS

For many reasons, when we taught middle school mathematics, we incorporated games (the old-fashioned kind, like card games, board games, trivia-type games) into the curriculum. We find that effective games enhance students' mathematics skills, change attitudes, and motivate students to learn. They build social skills, enhance problem-solving skills, help to minimize discipline problems, improve listening skills, and put some fun into the school routine. When you factor into the process how much time we can spend trying to challenge a student with worksheet-type review, we find the built-in motivating effect of games makes their use a time saver. Another positive outcome is providing opportunities for middle schoolers to socialize in positive and productive ways.

Katharine Owens,
The University of Akron,
Akron, OH

Richard Sanders and
Robert Lipsinki
Discovery Middle School,
Madison, AL

In the late 1980s, Kathie and Richard taught mathematics and science in the same geographical area for three years. Kathie, then a middle grades teacher, worked with Richard, then a middle grades teacher and later a member of university faculty, to develop a mathematics and science curriculum for middle grades. Kathie says, "It was apparent from the outset that we would work well together and had complementary strengths. We are both creative people; I am the verbal partner while he is the artistic one. His skill using computer software made it easy for him to design game boards and to prepare slides for presentations." Believing that games are wonderful mechanisms for motivating and teaching concepts or skills to students, they decided to collaborate on designing or modifying games to teach middle grades mathematics. "We both had similar groups of students with which to field test our ideas so we could easily make revisions in our games after we had both tried them with students," writes Kathie. However, a few years later, Kathie left middle grades to teach mathematics methods courses at a university about 400 miles away from Richard, and Richard moved out of the university and into a teaching position at Discovery Middle School. One would think that the distance would have curtailed their collaboration, but Kathie says, "because we are still committed to designing and sharing quality learning experiences for middle school students, we continue to work together on projects, although the time we have to work together is limited."

To prepare the materials for this unit, for example, Kathie and Richard spent spring break together creating, writing, and trying the games. They also spent time working with Bob Lipsinki, a colleague of Richard's, who teaches 6th grade mathematics, to prepare for Kathie's teaching of the lessons in Bob's class. Bob's class consists of students showing poor achievement levels on mathematics exams and having low motivation toward learning mathematics. These students are required to take a double period of mathematics to allow time for thorough review of concepts.

Days prior to class, Kathie and Richard share the strategy of the games with Bob, and they all create problems to reinforce topics he had previously taught. They also discuss Bob's teaching style and his use of games in his classroom. Kathie learns about any of his 6th grade students' special needs or characteristics, and about his classroom procedures—an important requirement whenever someone from a university setting comes into a middle school classroom to field-test new activities. Because the games are versatile enough to encompass the review of any number of concepts, the teachers have no problems adapting the games to topics previously taught by Bob. They look at the games to be sure the level of difficulty is appropriate for the students, and they discuss the management of the games. In keeping with Bob's preference for grouping students, they arrange the desks to accommodate groups of four girls or four boys.

ENGAGING STUDENTS

Upon entering class, students are surprised to see guests and desks arranged for group work because they are expecting an individualized test on this day. Students already know Richard, who serves as the school's media director, so Bob introduces Kathie as a university faculty member interested in working with him and Richard on creating innovative curriculum for middle grades. "Does that mean we won't get a test today?" a student asks eagerly. "Yes. You are in for a treat," Bob says. "Drs. Owens and Sanders have some games to play with you that will help prepare you for our test. They would also like to get your ideas on how effective these games are for your learning of mathematics." As students happily clear their desks, it is obvious to Richard and Kathie that they have a class of willing "guinea pigs."

Kathie distributes a sheet describing the goals and procedures for a game patterned after the game Dot-to-Dot (see Figures 10.1, p. 142, and 10.2, p. 143). Richard and Kathie's modification of this game places the focus on finding the factors of prime and composite numbers. Working in pairs, Kathie tells students, "Try to earn points by putting in the last stroke to complete a box. Now, here's how to earn points: You earn 5 points if you complete an empty box but points for a box containing numbers are determined by the sum of the factors of the number in the box. Let's think a bit about the boxed numbers. Which box would you try to box first, one containing 18 or 19?" After a minute, Julian raises his hand and says, "Factors of 19 are just 1 and 19, so the sum there is 20. But the factors of 18 are 18, 1, 6, 3, 9, and 2 so its sum is 39. I'd try to get the 18 boxed." "Good," she replies. "Note that Julian explains why boxing larger numbers may not always be the best strategy. Do think about which boxes will yield the greatest number of points?" One students raises his hand and asks, "Any prizes for winning?" "Interesting question," Richard says, "Let's agree that winners get a choice of Jolly Ranchers or M&Ms." Kids' faces light up further as they turn to the challenges of the game.

As they play, all three teachers circulate, make mental notes of students' interactions, and carefully observe for areas students might need help. Bob helps one of his students find a pencil and tells another to begin working. Kathie and Richard move among the pairs to answer questions and offer positive comments. Once students are playing smoothly, the teachers move to a side of the room and comment quietly to each other:

Kathie: Notice that table with the two boys. They are apparently very familiar with the game strategy because they were initially taking turns very rapidly. However, after a few minutes passed and the empty boxes got completed, they slowed down as they maneuvered their way to completing the numbered boxes. This is good because they are now really thinking and not just mechanically going through a process."

Richard: I noticed that at another table, two boys were struggling with picking out which numbers are primes and which are composites. I suggested that they look in their math textbook or journals for definitions and examples to guide them. That helped. I know that the game board contains some numbers that the class has already worked with and some new examples, so the boys' confusion shows that the game isn't just a matter of recalling past practice exercises.

Bob: Look at Claude. I call him the "clock watcher." But today, his complete attention is taken up by the game activity. By the way, Kathie, I know you are concerned about Leslie and Amanda, who seem to be off to a slow start. I usually have difficulty motivating these girls to participate. However, the work they are doing with this game is lots more than I usually get them to do without a struggle.

Kathie: I'll check on them once more.

Richard: Bob, check on Peter and Jake, too. Look at Peter's facial expressions. I think he needs help. What can you tell us about him?

Bob: I think I know the source of the problem. The two boys have different achievement levels. I should have grouped students whose achievement levels are more alike for this particular game. I'll see what I can do.

The teachers circulate again, and Kathie approaches the girls and asks if they are doing OK. Both girls, Leslie and Amanda, say that they like doing this work because it's not boring. Kathie is once again convinced that games are motivating influences and mechanisms for promoting learning. As Bob approaches Jake and Peter, he notices that Jake is winning many boxes while Peter is close to exploding because he hasn't had a turn in awhile. Bob moves closer to the pair, but says nothing at first. When the play finally moves to Peter, Bob urges him to take his time and to focus on a strategy before drawing his line. "There is no time limit for thinking, Peter," he says. That advice seems to help; a box is "won," and more balance between players is restored.

About 20 minutes have elapsed since the start of play. Pairs of students begin to finish completing boxes. Some students then exchange papers and proceed to check the factors of the numbers in the "win" lists and add up these factors to find the total points. The three teachers continue to circulate the room, assessing not only the game-playing process but also the accuracy of the students' work. Occasionally, one of the teachers asks a student to double-check a player's list of factors because the list might be incorrect or incomplete. To engage students who have finished early, Bob says, "For homework, we would like you to create a new factor game board for future games. Those who have already finished this game may start this homework right now." About 15 minutes later, when everyone is finished writing their factors and finding the sums, Kathie tells pairs of students to exchange their numbers list so that each student will compute each other's total scores. "Once you've computed the total score, please return the papers so that each of you can verify your own scores," Richard adds. The person in each pair who has the larger sum is the "winner" and receives a choice of Jolly Ranchers or M&Ms. The other person receives a lesser prize (a starlight mint), so *every student* has received some reward. The students pass in their papers so that Bob can record a participation grade in his grade book.

For the next game, Richard asks, "How many of you have played bingo?" All students' hands go up and he says, "Well this next game is played in the same way but with four big differences: First, rather than pull cards that have just a single number, we have created cards to help you review the concepts of number theory: odd and even numbers, prime numbers, multiples of 3 and 5, and rules for divisibility (see Figure 10.3, 144). Students groan a bit. This is not the bingo they expected! Henry asks "Dr. Sanders, why can't we just play regular old bingo?" As Richard draws the bingo square game board on the board (see Figure 10.4, p. 146), he tells Henry, "Well, it has all the excitement of the old bingo while requiring you to think. What are the rules for bingo?" "Well," Henry says, "When you fill in a row, column, diagonal, four corners, or the entire board, you call out 'bingo'. But what numbers do we put on the board game?" Kathie replies, "Use numbers from 1 to 100 and place them anywhere you want in a random way in the squares. The order in which you place the numbers is not important. Put one number in each box and do not repeat numbers." On the board, Richard quickly sketches a bingo square (Figure 10.4) and tells the students, "Here's a third difference: when a student correctly calls 'bingo.' The winner's array, that is, the row, column, diagonal, or four corners, now becomes closed to any other winner. The student who wins the first round becomes ineligible to win additional rounds, but may still keep playing along with the rest of the class in an effort to win blackout. Make a copy of the board. I will keep track of the bingo arrays won on the board so that you will know what possibilities are still open for play. Do not erase any of the previously crossed-out answers, because you should try to win a different row, column, diagonal or, aim for a blackout. On the overhead, we will show a list of the cards that will be called." "And guess what the fourth difference is?" Bob adds, "The prizes are different this time around. We have blow-pops for individual bingo winners, your choice of a special pencil or a one-night-of-no-homework coupon for blackout and peppermints for those who play, but do not win. So, everybody is a winner in our version of this game." "Now *that's* the kind of difference I like!" Henry comments.

FIGURE 10.1 DOT-TO-DOT BOARD

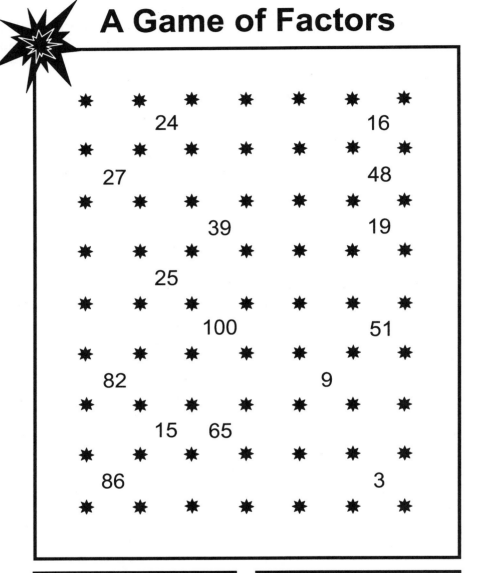

A Game of Factors

Player A	Player B
Name: _____	Name: _____
Number Factors	Number Factors

FIGURE 10.2 DOT-TO-DOT: THE FACTOR GAME

Description: The object of this game is to list the factors of composite numbers. Student partners compete against each other by completing squares with four straight lines on a grid of dots. As a student completes a square, he/she will write his/her initial in the square. The challenge of this game is to use strategies to win a square and to prevent the opponent from winning a square. This game is versatile—replace numbers with vocabulary terms, math problems, problem-solving situations, etc.

Number of Players: Two students per game with any number of pairs of students playing simultaneously. Since the game is competitive, students of comparable achievement levels work better together.

Teacher Preparation: Construct the game board as shown in Figure 10.1.

Directions for play:

1. Working in pairs, players (A and B) decide who goes first. Perhaps the person having the most letters in his/her first name could go first. Allow player 3 to 4 minutes to think of strategies.

2. Player A will connect two dots using a straight line up and down or across. Do not draw diagonal lines.

3. Player B will do the same. Player B's line does not have to connect with player A's line.

4. Player A takes another turn, and the turns rally back and forth.

5. The players continue connecting lines until a player completes a box. When a box is completed, that player writes his/her initial in the box and gets another turn.

6. A player receives 5 points for every empty initialed box but gets the sum of the factors of those numbers contained in boxes.

7. Points are calculated at the end of the game. The player who scores the highest number of points wins the game.

8. To calculate points, partners write their numbers list on a sheet of paper and then swap papers.

9. Using the opponent's list, players write all of the factors of each number on the list and calculate the opponent's total score.

10. Players return the list of numbers and verify the scores.

11. Teacher collects the lists of numbers and factors to inform later lessons and distributes prizes.

FIGURE 10.3 BINGO—THIRTEEN GAMES FOR THE
PRICE OF ONE: NUMBER THEORY VERSION

Description: One of the most popular games for reviewing the basic facts of arithmetic is bingo. Bingo makes a good game for review of content terms before tests. The rules are so simple that every student, regardless of ability, can easily participate. Knowledge of the basic facts is needed to win, but since an element of luck is involved, the perception exists that anyone, not just the best student, can win. In this version, students review concepts of number theory (odd and even numbers, prime numbers, multiples of 3 and 5, divisibility) using numbers from 1 to 100.

Number of Players: Students play individually. Teacher is the caller.

Preparation: Draw a 5×5 grid with 25 boxes on the board or overhead projector. Label the rows with letters A, B, C, D, and E and columns with letters V, W, X, Y, and Z. The students will use numbers from 1 to 100 to fill the 25 boxes. Put one number in each box in a random order. The teacher will make and cut out *four* sets of the cards listed below. Make an overhead transparency of the cards or write them on the board.

Sample Bingo Cards:

F. Odd number greater than 50

G. Prime number less than 50

H. Prime number greater than 50

I. Multiple of 5

J. Divisible by 3

K. Even number less than 25

L. Odd number

M. Even number greater than 50

N. Odd number less than 25

O. Even number

Directions for Play:

1. Teacher tells students to copy the bingo board on a sheet of player.

2. At random, the teacher selects a card, calls out the letter, and reads the corresponding description. Point to that description on the transparency or board.

3. If a student has a number on his/her bingo board that fits the description read, he/she writes the descriptive letter as it is called in the box and circles the number. Students can win any one of the rows, columns, diagonals, or "four corners."

4. When a row, column, diagonal, or four corners is completed, the student declares "bingo."

5. The winner must read back the letter called and the number that satisfies the description to verify a possible win.

6. The teacher or another student checks the accuracy of the answers circled against the cards called.

7. If the student's answers are incorrect, no "win" is declared and play continues as above.

8. If the student's answers match the questions called, the student is declared a winner.

9. A small prize, bonus points, or reward coupons may be given to the winner of this round.

10. The winner's array (row, column, diagonal, or four corners) now becomes closed to any other winner, but play continues with the teacher drawing and calling out more cards. Players do not erase any of their previously crossed-out answers, but use their crossed-out answers to build toward a possible win of a different row, column, diagonal, or four corners pattern.

11. The teacher keeps track of the bingo arrays won on the board, so that students will know what possibilities are "open" for winning.

12. The student who won the first round becomes ineligible to win additional rounds, but still keeps playing along with the rest of the class. Any student, regardless of array won, may win the board "blackout." A larger prize or reward could be given to the "blackout" winner.

FIGURE 10.4 BINGO BOARD

	V	W	X	Y	Z
A					
B					
C					
D					
E					

The students diligently plan their strategies to win a row or column. Several students have their math journals on hand to do some calculations or to check a definition. The game begins with Kathie calling, "J—multiple of 5," and pointing to those words on the overhead. Students having a multiple of 5 on the board, circle one such number and put the letter J next to it. She next calls, "H—prime number greater than 50." Before Kathie calls each card, she surveys the room to see if everyone is ready. She is careful to make adjustments for each student's pace and knows how important good listening skills are for this game. She also arranges the cards neatly on her desk so that she will be able to match the cards to students' bingo calls. Bob and Richard, meanwhile, notice how Jason is trying to change the numbers on his card so he can make a bingo. Making eye contact with Jason, Bob tells the class, "Remember, once you've placed your numbers, you are not to change them. That's called cheating in our version. Players changing numbers will forfeit the game and lose participation credit." Jason gets the message—from then on he follows the rules and eventually honestly wins a bingo row.

Again, the teachers assess students. They note that students had no difficulty marking even numbers as composite numbers, but some of the odd numbers posed more of a challenge. Bob observes that Farah needs some help with deciding whether 51 is a prime number. Bob asks her to recall the tests for divisibility, which previously had been taught. As soon as Farah remembers the "test for three," she knows that 51 is not a prime number. Noticing that quite a few students need similar help, and as result are incorrectly calling bingo, Bob asks Kathie to consider stopping the game midway to review these divisibility rules with the whole class. That was a good move because from that time on, only one or two students had a wrong answer when they called bingo.

An interesting question arises in the course of play. Beth could not remember if the number 1 is prime or composite. Kathie, knowing how flexible middle school teachers have to be and how important it is not to miss a teachable moment, asks the class for opinions. Playing the game is suspended while students discuss and seek to answer Beth's question. Claude checks his math book and reads out loud, "The number 1 is neither prime nor composite." "Who can tell us why it makes sense to say that?" Richard asks. After a three-second pause with no hands raised, Richard adds, "Go back and look at the definition for prime and composite number." Looking at her book, Tiffany raises her hand to say, "The number 1 has only itself as a factor. Composites have more than two factors and primes have exactly two factors. So, 1 doesn't fit either definition." "Good," Richard replies, "now let's resume playing."

Wilson and Brian call bingo at the same time, and both have covered numbers along the second row. Kathie verifies their correct answers and rewards both boys with a prize. Finally, only one more way remains to be won on the boards. The teachers feel the tension in the room as those who haven't won yet double their concentration, hoping for, and at the same time dreading, that next call of bingo. When Tawana calls "bingo!" a sigh of disappointment can be heard. "Call out your numbers," Kathie says. As two students look over Tawana's shoulders, she says, "Going down the V column, I have: F-53, J-35, P-21, Q-14, and P-63." As she calls her cards, Kathy looks at the cards on her desk for verification, and before she asks the class to decide if Tawana has indeed made bingo, Amanda's hand shoots up while she says, "Hold it! 63 is not a prime number!" "Tawana, do you agree?" Kathie asks. "Ummm...Yeah. Nine times seven is 63, so it is composite," she replies disappointedly. A sigh of relief is heard from many students, and the play resumes again.

Eventually, the last bingo possibility is won and everyone works toward "blackout." It takes only five or six more calls before Kayla shouts "bingo." Two students verify her bingo and she chooses the free homework pass as her prize. Because of multiple correct "wins" called at the same time, 18 out of 26 students have won blowpops and one has won the bigger prize. Students who didn't win anything are given a peppermint so that everybody is enjoying some treat. Bob tells his students to pass in their bingo boards so he can check their work for accuracy and completeness.

Interested in getting feedback from students, Kathie says, "Please share with us your opinions or ideas about the games we have played today." Several raised hands indicate that these "guinea pigs" have lots to say. "This was a fun way to learn," says Wanda, and her partner nods her head in agreement. "I liked this better as a way to get ready for our test than doing a worksheet. I learned a lot. It was cool," remarks Ashley. Claude, "the clock watcher," comments on how quickly the math class had gone by and couldn't believe it was almost time to change classes. Richard asks, "How did you feel when you didn't win?" Jason offers, "It was just a game; I didn't feel bad, although I really wanted to win. I'll win the next one. Jennifer remarks, "I almost won that first game; I'll bet I can win the second one." Kathie is quietly cheering. Kathie then asked the students for their opinions on the bingo game. "I learned a lot and had to review what we did in the book to be able to play," remarks Steven. Cory admits, "I think it was hard to keep up sometime and I think I missed calling bingo one time because I was behind." Bob later reports that Cory struggles in math and could probably benefit from an oral review of the concepts before the game begins next time. Tawana confesses that she was frustrated when she didn't win, but liked playing nonetheless. Leslie and Amanda ask Bob if the class can play again sometime. Bob tells them if they make up different calling cards then they'll have it to play before the next chapter test. The factor game boards from tonight's homework will also be used. Angie wonders if they can play bingo in social studies, too. Richard offers to share the rules and procedures with the social studies teacher, because this game is very adaptable and would be good for reviewing vocabulary words.

Several students asked Kathie to come back again and play some more games. In fact, Jason asked Kathie and Richard if they had any other games, because "your games are cool" and "this was fun." At that point, Kathie and Richard wished they were back in the middle school classroom full time! They too had fun.

DISCUSSION BETWEEN COLLEAGUES

What were your impressions of girls' versus boys' participation?

> *Bob:* I saw girls who do not normally participate get actively involved in all games. For example, one of the girls who generally sits back and needs pushing to get involved really got motivated and excited about playing.

> *Kathie and Richard:* Participation was just about the same for boys and girls throughout the class. We noticed that when a girl won a round of bingo, she tended to be more timid in calling out her answers for verification. For example, one girl had a "false win" (one of her bingo answers was incorrect), but in the next call, she did get bingo. She was very shy about calling out her winning series of numbers, but very proud when she was right this next time. We got more feedback from

the girls when we asked for a critique of each game. This surprised us somewhat; we thought we'd hear from at least an equal number of boys as girls.

Are there characteristics of the games that make them more suitable to one learning style than another, and if so, any suggestions?

> *Richard:* I did not notice any specific thing. Kathie is an outstanding teacher who monitors and adjusts—moment by moment. She uses a variety of learning styles that attract quiet girls and troublesome boys.
>
> *Kathie:* Games imply competition, which, for some students (shy boys and some girls), can be "unfriendly." To diminish the anxiety some students would feel, we manage carefully and apply our assessment of the students as the game progresses. Games with multiple opportunities to win and many winners within the same game, or an element of chance that inspires even weak students to think they have an equal chance to win as do the smarter students, help take the "edge" off the potentially unfriendly competition. I've tried to use a great variety of games over the years—some are for partners (like the dot-to-dot game), some for small groups (like card games), and some for larger groups (like blackboard relays). Some games are geared toward visual learners (like Concentration); some require writing answers others require good listening; some require motor skills (even board games can be tough for little ones when they must move little markers around the track); all require thinking skills and forming a strategy for winning; and all require use of social skills, fairness, logic, cooperation, being a good sport, being a good loser as well as a good winner. Regardless of their learning style, every student can benefit from the opportunity a game affords to improve decision-making skills.

Comment on the prizes: Were the kids playing just for the prizes?

> *Bob:* The prizes weren't necessary. I am sure the kids would have played for free.
>
> *Kathie and Richard:* We are glad we had small prizes because we think they improve motivation. We also like to keep the size of the prizes in agreement with the size of the task; for example, making the "blackout" prize something larger, more unique, or more special than the prizes for individual rows and columns. We have used lots of food prizes in the past (middle schoolers are always hungry) and have had

some backups for kids who can't have chocolate, sugar, etc. We've used special pencils or other school supplies, tokens to save for larger prizes, and money (our own; only for really BIG games); but other good prizes could be free-time, extra-recess, and homework passes or other privileges.

What were your general impressions on using someone else's class to test your ideas?

> *Richard:* I did not have any negative feelings about using students from my school. They are great kids and I like to work with them.

> *Kathie:* I was a bit concerned that my lack of knowledge of Bob and of his kids (I like to know more about the abilities and personalities of the kids I work with, for example) might make for some rough spots in the day, but that really never happened. Bob has great rapport with his students and I come across as a friendly, middle school teacher (not a pompous university professor), so the tone in the class was quite comfortable.

Bob, what are your impressions of having Richard and Kathie take over your class?

> I would welcome them again! I learned a lot about using games and would use all of the games again. The games, especially bingo, could be adapted to whatever skill I wanted to review with students. I also saw how easily the games motivated my students to work on a very rough day for teacher and students alike—the Friday before spring break—and yet the students were on task the whole time. I was also impressed at the relatively low noise level in the room. I've played games before in his classroom, and sometimes the noise level gets out of hand. The only real noise I heard was when students cheered their completion of several boxes in one turn. However, this occasional cheer seemed to spur other groups to achieve similar accomplishments. It was also fun for me to see how much fun the students and teachers were having. These games had the students engaged in a challenging task, which they eagerly carried out. I feel fortunate to have worked with Kathie and Richard because I now have fresh ideas to share with others teachers.

Any snags?

> *Kathie:* I would have liked more background on the skill level of the students, so I could make a better judgment about the kids' learning through the games. Also, it was best that I follow Bob's procedure for

groups. What works best for him is to group his kids so that boys work with boys, girls work with girls; I like to mix the genders in groups and give definite roles to group members, switching periodically so that everyone gets leadership jobs. However, Bob's presence in the room helped the process go well.

What would you do differently next time?

Bob: I'd do a quick practice game because some students may need them to understand fully what will be expected during the game as well as a review of the underlying concepts.

Richard: Each group is different. I would not change any one specific thing—but monitor the next group and make adjustments based on what I saw. For example, there are always students who will try to cheat. When I am the only teacher, I tell students that one other student is to supervise the calling of another student's bingo.

What procedures do you use to help students make their own games?

Kathie and Richard: We find it best to assign pairs of students the task of making a game modeled after popular board games. For example, to emphasize the meanings of chapter vocabulary, we have students make and play board games (like Sorry), trivia games (like Jeopardy), or card games (like Old Maid, Concentration, or Dominoes). We have found the game format for "teaching" vocabulary terms to be especially useful when working with the reluctant reader who then gets motivated to read the text, reword the meanings of the terms, and use these terms in the competition. We made handouts for the kids to give guidance for making a game. Tips for making rules, game components, and peer review sheets are in the handouts. Of course, we then have a game-playing day when students swap games and play for modest prizes. A note on classroom environment: We emphasize to students that "putdowns" will not be tolerated, and if these occur, fun activities will immediately stop; Our overarching goal is to orchestrate learning opportunities for students to feel successful overall while simultaneously providing a safe, supportive environment for them to practice being good losers.

What factors influence your strategy for modifying games?

We strive to make a good variation that will reinforce important concepts while allowing for lots of flexibility, many winners, and minimum equipment. We create cooperative games, too, since competition is more male-gender-friendly. We also like to include a chance factor in our games because even a less-capable student feels that he or she has a chance of being successful when the game involves not only knowledge but also luck, as in the roll of a dice, the draw of a card, or the spin of a wheel. Bingo, for example, has been a mainstay for review in classrooms for decades, but our version maximizes the number of winners and avoids the mess of "bingo markers" scattered about the room. Once someone wins, the other students play even more eagerly, hoping to be the next winner. Also, because they're striving to get blackout, everybody stays engaged until the end of play.

Richard, as a media resource person, how do you continue to help teachers and others implement standards-based teaching?

My role in helping the mathematics teachers varies. My main job involves working with media resources—that includes the library and classroom equipment—and helping students develop research skills, helping students develop the reading habit, and training teachers and students to use the Alabama Virtual library, along with other Internet sites. I have a very extensive personal collection (estimated at $30,000, the product of 20 years of collecting) of teacher education, social studies, mathematics, and science teacher activities books at home. The school's teachers know about my personal library and ask for specific help when they need it. Sometimes I may be asked to mentor or help a teacher in any subject area —I do so. New teachers and I spend lots of time together for media orientation. I help most new teachers with equipment, materials for teaching, and unit planning.

COMMENTARY

Johnson (NCTM, 1975) writes, "Unquestionably, the key to learning arithmetic is through meaningful experiences and practical applications. However, the skills of computation need to be nurtured by a variety of systematic drill and practice....Arithmetic games are ideally suited for a systematic program of practice" (5). It's interesting to note that his article in the NCTM collection of games for elementary and middle school mathematics was first written for NCTM in 1958. However, Johnson's message about arithmetic and the use of games, then and today, is supported by NCTM. Johnson states that the success of a game is dependent on how it is used. "If an arithmetic game is to serve a real function, it should be used at the right time, for the right purpose, in the right way" (5). He gives five recommendations for using arithmetic games:

♦ Select a game according to the needs of the class—the game should involve important concepts or skills and be closely related to the work being done in class.

♦ Use the game at the proper time—best times are when the concepts embedded in the game are being taught or reviewed in class.

♦ Arrange the game situation so that *all* pupils will be participating in every play—also, avoid embarrassing anyone. Keep the learning environment safe and pleasant.

♦ Plan and organize the game carefully so that the informality and excitement of the setting does not defeat its purpose—pupils may referee as well as play. Choose group members carefully so that low-ability students are not embarrassed.

♦ Emphasize the responsibility of learning something from the game —follow-up discussions or tests help stress a game's importance to learning. Teachers should also assess the success of the game based on its impact on student's learning of desired concepts.

According to the National Science Teachers Association (NSTA 1977), games are more effective at changing attitudes than any other teaching technique. When used skillfully, games can build social skills, self-esteem, and confidence. When the game format challenges students to make choices and use a strategy to win, students' problem-solving skills and critical-thinking skills are enhanced. What's more, curriculum objectives are met in fun ways when the game is not just a "time filler." Playing games that involve both good questions on important mathematics and effective group participation are key features in reducing the boredom and idleness too often felt by classroom students. Thus, the use of such games can minimize some of the discipline problems in the middle school classroom. Games also help students develop important social skills, because the students know they need to listen carefully and pay attention to the directions of the teacher, to the course of play, and to everyone's answers.

Kathie and Richard's use of games incorporates Johnson's recommendation and makes all of the NSTA's characteristics of effective games happen in the classroom. The concepts for the games are chosen to coincide with those being taught in class. Students are engaged because they are curious about the answers, feel they have a good chance to win, are eager to test or display their knowledge, and are eager for the prizes!

Many recommendations from the *Standards* are promoted in this lesson. Kathie and Richard's success in making games for learning some concepts of number theory, which include multiple answers and an element of chance, gives all students a chance to feel that they can be winners in mathematics. (number and operations, equity, problem solving, communication). For teachers, Richard and Kathie show how games can be used to address specific learning objectives when planned wisely. They show the use of games as a coherent part of the curriculum (curriculum principle), rather than as a five-minute, end-of-class activity Their teaching of these games in Bob's classroom also strongly models a collaboration that fosters professional development through teachers and university faculty working together to develop challenging and appropriate lessons for students (teaching principle). The students were challenged to review or learn important concepts from games that combined both factual knowledge and conceptual understanding (learning principle). Finally, based on students' answers as the teachers circulated among the groups, the teachers assessed areas where students' needed additional help and, when necessary, redirected the lesson to meet those needs (assessment principle).

For readers interested in seeing a game devised by Kathie and Richard for middle-grade students' review of place-value concepts, computation of whole numbers, and large number values and number names, see Owens and Sanders (1992). For an article that includes a good database of commercial games correlated with the NCTM Standards embedded in the game, see Leonard and Tracy (1993).

CONTACT

Kathie Owens
The University of Akron
130 Zook Hall
Akron, OH 44325-4205
E-mail: kowens@uakron.edu
Phone: (330) 972-7437

Richard Sanders and Robert Lipsinki
Discovery Middle School
Madison, AL 35758
E-mail: richardsanders2@aol.com
(256) 837-4104

UNIT OVERVIEW:
GAMES FOR PRACTICE AND
REINFORCEMENT OF CONCEPTS

Aim: How can games help us better understand factors and divisibility of numbers?

Objectives: Students will play games to reinforce the concepts of primes, composites, and the rules of divisibility.

Grade Levels: 5–6

Source: Original

Number of 90-Minute Periods: 1

Mathematics Principles and Standards Assessed:

- Number and operations

- Mathematics as problem solving, communication, and reasoning

- The principles for equity, curriculum, teaching, learning, and assessment

Prerequisites: Prior instruction on the concepts of prime numbers, composite numbers, and divisibility rules 2, 3, and 5.

Mathematical Concepts: A review of he concepts of prime, composite numbers, and divisibility rules 2, 3, and 5

Materials and Tools:

- Instructions for the games (Figures 10.2, p. 143, and 10.3, p. 144)

- Game boards for games (Figures 10.1, p. 142, and 10.4, p. 146)

- Prizes for winners

Management Procedures:

- For the factor game, arrange students in pairs of similar achievement levels and refer to Figures 10.1 (p. 142) and 10.2 (p. 143).

- The bingo game is for individuals. Refer to Figures 10.3 (p. 144) and 10.4 (p. 146) for instructions.

Assessment: Circulate around the room as students play. Consider reviewing concepts that cause students to incorrectly call bingo. Have students assess each other's work and check papers afterwards. Consider giving participation points for completing game.

11

PATERSON SCHOOL 2: JOURNEY BEYOND TIMSS

In 1991, Paterson Public Schools were taken over by state, All the schools were assessed and our PreK-8 School was declared one of the four worst in the system. Our population of 720 students is challenged by many factors: Ninety-eight percent of our students qualify for free lunch; many come from high-stress environments, including a hotel for the homeless, a battered women's shelter, and a housing project for children with seriously disabled parents; ten classes of special education students are bussed to the school from throughout the district; and our rate of transience is 42%.

Today, collaborative support for applying the Lesson Study process of Japanese teachers in our own classes has helped us create a study group to focus on enhancing the teaching and learning of mathematics. Four main areas guide discussion of the group: curriculum, instruction, professional development, and school culture.

> Middle School Lesson Study Group team members from Paterson School 2, Paterson, NJ: Lynn Liptak, Fran Dransfield, Bill Jackson, Isabel Lopez, Magnolia Montilla, Beverly Pikema, Cynthia Sanchez, and Nick Timpone

On April 13, 2000, I (Yvelyne, the author) first learned of the work taking place at Paterson Public School 2 during a national meeting of the Middle School Mathematics Professional Development Network, which is a project of the Eisenhower Regional Consortia and National Clearing House. My very first question to Bill Jackson, who has taught at School 2 for 16 years and serves as facilitator for mathematics professional development in the school, was, "Why try to import the Japanese way of doing things? We are so different." His response was simple and clear: "What we were doing was not working for our students. Our students were failing miserably, so we had nothing to lose. It's the TIMSS [Third International Mathematics and Science Study; see Chapter 1] videotapes that really grabbed out attention and interest. We saw ourselves teaching in the same way as the American teachers who offered no challenges for students to learn the underlying concepts of mathematics. We wanted to give our students opportunities to enjoy rich mathematics through challenges." Because of my interest in profiling the study group teachers work for this book, Bill invited me to an Association of Mathematics Teachers of New Jersey (AMTNJ) conference where I'd see lesson study in action and get an opportunity to meet the major collaborators assisting the teachers.

The conference was held at School 2 and sponsored by AMTNJ, School 2, the Mid-Atlantic Eisenhower Consortium/Research for Better Schools, Columbia University Teachers College, and the Greenwich Japanese School. Because of this extensive collaboration and the unique approach to curriculum taken for helping School 2 teachers strive to improve students' learning of mathematics, this profile will stray from the other profiles in this book in that I've included a lot of my own personal comments or reflections as I witnessed this novel, intriguing, and powerful form of professional development. Three other differences are that the Unit Overview Plan is a section from the actual lesson plan used by the teachers and is placed *before* the Commentary section; all references cited, as well as additional sources of information on the school and study group process are also compiled at the end of the chapter (Figure 11.3, p. 187); the lesson profiled is at the 5th grade level because it is the one I had the opportunity to observe. A plea to readers: Do resist the temptation to skip over these details and to rush to the classroom lesson. You may miss the impact of a major point I wish to make about the process. Patience, please.

THE JOURNEY

Now, the journey beyond TIMSS begins. As I approach the school, I notice that it is nestled in the heart of Paterson's downtown, a heavily traffic-congested area. Pretty dreary, I think. But when I enter the school, I soon forget about the outside because School 2 beams with a generally warm and welcoming atmosphere conveyed by its administrators, teachers, and students. Nice bulletin boards, good lighting, and a clean space all enhance this atmosphere.

As my luck would have it (it's always good), I sit beside Frank Smith, a Columbia University Teachers College professor who first offered workshops for the principal and teachers when the State Department of Education "took over" the school. "What does it mean for a school to be taken over by the state?" I ask

Frank. He explains that the district of the school is declared low performing and mismanaged. The state disbands the board of education, replaces the superintendent, reassigns or removes the principal, and replaces the central office staff. At School 2, the principal was replaced with Lynn Liptak who, in 1996, became interested in Japanese teaching methods after attending Frank's workshop on the design and results of the TIMSS study (see Chapter 1 for information on TIMSS).

Lynn summarizes what happened next, "The TIMSS 8th grade videotape study provided a powerful contrast between United States and Japanese mathematics instruction. We decided to try it and were encouraged by responses of our students when 'Japanese-style' lessons were taught. Groups of School 2 teachers spent the next three summers writing and revising 7th and 8th grade lessons aligned with the New Jersey Core Curriculum and influenced by TIMSS." In 1999, Lynn decided to try the Japanese approach to professional development through lesson study because, she says, "I don't believe in having teachers work all day and then having to attend professional meetings after school. Such meetings are best conducted as integral parts of the school week." Lynn schedules a study group once a week for teachers interested in innovative curricula.

What is this "Japanese style," and what help did teachers receive to learn and implement it? This question was clearly addressed throughout the conference. Catherine Lewis, an educator from Mills College, Pennslyvania, spoke to the question by drawing from both her vast research on Japan's educational system (Lewis 1995) and her actual videotapes of the Japanese process. Excerpts from one of her articles (Lewis 2000), describing lesson study follows.

LESSON STUDY

Research lessons or (study lesson) refers to the lessons that teachers jointly plan, observe and discuss....Research lessons are actual classroom lessons with students, but typically share five characteristics:

1. Research lessons are *observed* by other teachers. The observing teachers may include just the faculty within the school, or a wider group; some research lessons are open to teachers from all over Japan.

2. Research lessons are *planned* for a long time, usually collaboratively.

3. Research lessons are *designed* to bring to life in a lesson a particular goal or vision of education. The whole faculty chooses a research team or focus.

4. Research lessons are *recorded*. Usually teachers record these lessons in multiple ways, including videotape, audiotape, observational notes, and copies of student work.

5. Research lessons are *discussed*.

A colloquium follows the lesson. Typically, such a gathering begins with presentations by the teachers who taught and coplanned the lesson, followed by free or structured discussions, and sometimes an outside educator or researcher also comments on the lesson.

This process is also widespread in China, but how could it work in the U.S.? I can list numerous reasons why some would believe that it has little or no chance for success in the U.S. In his *Los Angeles Times* article on teaching techniques, Richard Cooper (1999) interviewed a number of educators for perspectives on differences in the cultures of American and Japanese teachers that have bearings on teaching and learning. I use some of Cooper's sources, as wells as others, to find some reasons why one might argue that lesson study should fail in the U.S.

TOP TEN REASONS WHY LESSON STUDY SHOULD FAIL IN THE UNITED STATES

1. Stigler and Hiebert (1999, 99): Systems of teaching are much more than the things a teacher does. They include the physical setting of the classroom; the goals of the teacher; the materials, including textbooks and district or state objectives; the roles played by the students; the way the school day is scheduled; and other factors that influence how teachers teach. Changing any one of these individual features is unlikely to have intended effect.

2. Stevenson and Hiebert (1999, 157): One of the biggest challenges schools will face is that there are few leaders among its teachers for launching this process. Very few teachers have experienced this kind of professional development.

3. Stevenson and Stigler (1992, 157): Americans often act as if good teachers are born, not made. We hear comments implying this from both teachers and parents. They seem to believe that good teaching happens if the teacher has a knack with children and keeps them reasonably attentive and enthusiastic about learning.

4. Eugene C. Schaffer, University of North Carolina (Cooper 1999): In the culture of American education, you don't come into my classroom and I don't come into yours. That's a long tradition in this country. It's another element of American individualism.

5. Cooper (1999): Even when teachers work together in teams, as many do in middle schools, and time is budgeted for collaborative planning, the focus is often not on improving specific lessons. One recent survey found that middle school teaching teams often spent most of their meeting time discussing discipline and logistics for things like field trips instead of instruction.

6. Lenz, Deshler, and Schumaker (1990): Many (American) teachers report that their most productive planning occurs in unstructured settings (e.g., driving a car, walking the dog, or taking a shower). Teachers generally agree that during-school planning involves "top structure" planning (e.g., scheduling, determining class structure, and responding to administrative tasks), whereas "deep structure" planning (e.g., how to clarify a difficult concept or develop a motivating activity) is accomplished away from school.

7. Catherine Lewis, Mills College (Lewis 2000, 16): Japanese teachers spend little time…in developing or aligning curriculum, or translating national standards into practice. They have a frugal course of study and a number of nationally approved textbooks from which to choose…elementary textbooks are written by elementary teachers, based on their actual lessons. Because Japanese teachers start with texts that are teacher-written and lesson-based, they can afford to spend considerable time…planning, observing, and discussing actual classroom lessons.

8. Marjorie Coeyman (2000a), *Christian Science Monitor*: U.S. teachers have developed a thick skin against frequent reforms that encourage wild swings or suggest a lost golden age of learning. They're not helped in their day-to-day efforts by a culture that prizes reading far above math in early grades, and where kids can define "nerd" well before they learn to multiply.

9. Patsy Wang-Iverson, Research for Better Schools (Hoff 2000): The limitation in the United States is the lack of support for teachers to grapple together with the teaching of mathematics so it makes sense to the students.

10. Readers, please add your own reasons.

Yet, I learned that in 1998, and similarly in 1999, while School 2's 8th grade students' scores increased by 20 percent, their pass rate increased to 77 percent. This is only one year after using a problem-solving approach to produce lessons. Given the statewide pass rate of 86 percent, that's pretty darn good! "While we are proud of the results, we warn readers not to assume that Japanese methods were the reasons for the increase. There are too many variables to consider," cautions Lynn. Stigler and Hiebert (1999) write, "For lesson study to be a viable means of improving teaching nationwide,...two tests must be passed. First, lesson study must meet the needs of teachers....Second, lesson study...must meet the needs of the U.S. education system. Teachers are under great pressures to perform and the stakes are getting higher" (151). With so many strong arguments against its success, how could lesson study be active at School 2?

SIX REASONS WHY LESSON STUDY IS WORKING AT SCHOOL 2

1. The principal, Lynn Liptak, and vice principal, Fran Dransfield, both support the process. For Lynn, the combination of experiences provided by lesson study is better than any seminar the teachers attend outside the school. "They're really talking about teaching and learning," she says, "They're talking about what happens in the classroom. That is professional development right where it belongs—in the classroom—driven by teachers." In addition to her administration role, Fran is herself one of the study group teachers.

2. Because the teachers are part of the process, they are motivated to change the way they teach in an effort to focus on conceptual understanding.

3. The teachers have teams collaborate to do the work. Over the past three summers, Bill Jackson, Beverly Piekma, Nick Timpone, and Magnolia Montilla have earned stipends to develop the curriculum and write lesson plans to share, discuss, and revise with the other teachers. In his role as math facilitator at the K-8 school, Bill runs math meeting periods once a week per grade level. Also held every week is a lesson study period where teachers from all grade levels gather to share their ideas of how they will write or revise lessons that encourage in-depth thinking.

4. The Lesson Study Research Group (LSRG), based at Columbia University, Teachers College and directed by Drs. Clea Fernadez and Makoto Yoshida, helps the teachers understand the study group process and documents the work. The goal of the LSRG is to provide careful research about lesson study and how it can be adapted to the U.S. context.

5. The Japanese School of Greenwich, with the assistance of it's principal, Mr. Koichi Tanaka, provides help and guidance to School 2.

Opportunities for both groups to observe each other conduct a lesson study are part of this unique collaboration.

6. The Regional Eisenhower Consortium for Mathematics and Science Education, a nonprofit organization, provides funds for professional development of the teachers as well as the assistance of Dr. Patsy Wang-Iverson, who is its senior associate for Research for Better Schools.

Specifically, how did Public School 2 teachers apply the lesson study process to make it work for them and their students? What do teachers *actually* do during these meetings? LSRG gave a handout that succinctly defined lesson study and listed four steps summarizing the process used at by the teachers (Figure 11.1). Below each of the four main activities listed, School 2 teachers describe how they applied it.

FIGURE 11.1 LESSON STUDY AT PS 2

"Lesson study" is a Japanese approach to teacher professional development that involves a group of teachers working on four main activities:

1. *Setting a goal they all want to achieve with their students.* In our goal to implement lessons that develop "profound understanding of fundamental mathematics" (Ma 1999), we have become aware that our students need to take more responsibility for their own learning. Therefore, we strive to provide instruction that emphasizes problem solving, logical thinking, and student autonomy. We also try to provide instruction that actively engages and motivates our students to learn.

2. *Planning a lesson study (with a detailed lesson plan), which they will use to examine their chosen goal.* We meet once a week for two hours and plan in groups of four or five to do the work. We spend a lot of time thinking about the development of the lesson because we moved from lesson plans that were hastily written to cover four components—objectives, procedure, material, and homework—to lessons requiring thoughts on what sort of questions should we ask and how might students respond? How should we pace the development? What about individual differences? How should we pace the development of the ideas? (We learned how to pace by observing our Japanese colleagues because they are constantly aware of time.) How can we tell if students are motivated or have learned the concept? What materials can best help students understand the main ideas? The material section required a good portion of time for

making or finding appropriate materials to enhance the ideas of the lesson.

The process also encourages good discussions on the mathematics content and thus helps us to better understand the mathematics in the lesson. Some of us have more experience teaching than others, or a keener understanding of the underlying mathematics. We appreciate the help we receive from each other and the opportunity to share ideas that are later refined and improved upon by the group process.

Once our brainstorming is clarified, deciding who does what just falls into place: Those good at typing take notes or create worksheets while others work on making manipulatives.

3. *Teaching the lesson study in a real classroom while other group members observe.* At first, it felt strange having five or six other teachers in the classroom, but we soon got used it. It's nice having colleagues in the class observing the process because they see things to improve learning that one teacher may not see. The teachers also provide immediate feedback on the progress of the lesson to help guide the lead's teacher next steps. However, we are sometimes guilty of doing too much and must remember to be careful not to interfere with the progress of the lesson. The students wondered about this process but they always rise to the occasion because they like attention and want to feel important.

4. *Debriefing to reflect on the instruction witnessed and discuss what it taught them about the goal they set out to explore.* Before this process, we taught a lesson, reflected very little on the outcome of student learning, and then moved on to the next. Now we take the time to check that students are understanding the concept. Knowing that we have to start the discussion with our own reflection on the lesson causes us to be more alert to students' comments and reactions. It also a humbling experience because there is always something that we could have improved. What makes the process so worthwhile, however, is knowing that comments or criticism we receive bring us one step closer to improving our students' understanding of mathematics and our own understanding. Indeed, if it is perfect, then we have learned just a little.

Having shared comments, we next revise the lesson, teach it again to a different group of students, debrief, and eventually find ways to disseminate it to other teachers.

A Profile of Japanese-Style Teaching at School 2

We are finally at the point of our journey where we visit the classroom. Figure 11.2 shows a section of the actual lesson plan created by study group members Cynthia Sanchez (lead teacher), Fran Dransfield (vice principal), Magnolia Montilla, and Nick Timpone. The section reflects the process used to create the entire lesson. My thoughts, as I watched the process, are in italics throughout the episode. As you read the classroom episode, pretend to be an observer whose comments will be solicited later for the colloquium portion. As such, you should know that there are guidelines to follow. Figure 11.2 (p. 169) is a conference handout of the guidelines for observation. Keep these in mind as you "observe" the lesson.

Unit Overview: A Section of the Fifth Grade Mathematics Lesson-Study Plan

Teacher: Cynthia Sanchez

Writers: Cynthia Sanchez, Fran Dransfield, Magnolia Montilla, Nick Timpone

1. **Name of the Unit:** Circles

2. **Instruction Plan:** 5 lessons

   - Parts of a Circle

   - Discovering Pi (present lesson 5/22/00)

   - Circumference of a Circle

   - Area of Circles (2 lessons)

3. **Instruction of this Lesson**

   Lesson title: Discovering the Meaning of Pi

   Goal of the lesson: Following a lesson on the parts of a circle, the students will use the parts of the circle to discover the relationship between the circumference and diameter of a circle and the number pi. Students will work with a variety of circles to establish a pattern that will show that the relationship among the circumference and diameter of a circle and the number pi are constants for all circles.

   The relationship of the goal of the lesson to the goal of the subject: We would like to have the students gain a solid understanding of what the number pi stands for and how it is derived. The lesson will require students to interact with one another to develop a method for measuring the circumference of a circle when all that they know is the diameter of the circle. Students will be responsible for using their prior knowledge and the input of their peers to solve problems with little support from the teacher. Class will have to work as a whole group at end of lesson to discover that the relationship between the circumference of a circle and pi is a constant that works for all circles.

Materials: Video of teacher riding on a Ferris wheel, large oak tag posters representing the Ferris wheel for each group, individual worksheets with circles and chart for recording information, large chart for black board, large drawing of the pi symbol, circle and diameter manipulative, and basket at each table with string, markers, scissors, tape, and calculators.

Step	Student Activities	Teacher's Support and Things to Remember	Evaluation
Intro., 5 min.	Students will listen actively and recall information from previous lessons.	Discuss previous lesson. Elicit responses from student's regarding the parts of a circle. Put terms on the board: circle, radius, diameter, circumference. Link review to today's lesson. Today we will be learning more about the circumference of the circle. Define circumference again and introduce the video.	Were the students able to recall information from the previous lesson? Were students motivated by the video?

Step	Student Activities	Teacher's Support and Things to Remember	Evaluation
Pose the problem and hand out materials, 5 min.	Students will listen actively for instructions and will ask questions about the problem. Students may ask what one complete time around means.	You saw me (teacher) on a Ferris wheel. I started at one spot and went around and around. Let the Ferris wheel be a circle and the distance from the bottom of the Ferris wheel to the top of the Ferris wheel (100') be the diameter of the circle. Can you figure out how far I traveled when I went once around the circle when all that you know is that the diameter of the circle is 100'? Illustrate the problem with the student worksheet. Tell students to think about how they might solve the problem. Take suggestions and list them on the board. Hand out to each group a large poster board with a drawing of the Ferris wheel problem. Each group should also have a basket of supplies. Remind students to answer the question: How many times does the diameter go into the circumference? How did you find out?	Are the students asking questions about the problem? Do they understand what they are being asked to do? Are they offering suggestions for solving the problem? Are they anxious to begin working on the problem?

Step	Student Activities	Teacher's Support and Things to Remember	Evaluation
Problem solving, 10 min.	Students will work in their groups to find a number of different ways to solve the problem. Anticipated responses: Students may use string to try and measure the circumference of the circle. Students may redraw the diameter in a few different places and add them to find the circumference. Students may use string to measure the circumference and compare it to the diameter.	Teacher will move around the room looking for various and different solutions to be presented after problem solving is completed.	Are the students communicating with each other in their groups? Are they using more than one method to solve the problem? Are they using prior knowledge?
Present solutions, 15 min.	One student from 3 or 4 groups will go to the front of the room to display, present, and explain their solution to the problem. Students not presenting will be listening and questioning.	Teacher will call on specific groups to present. Presentations should be varied and presented in order of least correct to most correct. Teacher will encourage students to ask questions of each other and to debate the solutions.	Are the students, who are presenting, speaking in a loud clear voice? Are their explanations clear and understood by the class? Are the students questioning and debating?
Summary, 3 min.	Students will listen actively to teacher's summary.	Summarize student's solutions using presentations on the board. Make connections between different solution methods.	Do the students see the merits of the different solution methods? Do the students see the relationship between the circumference and the diameter of the circles?

FIGURE 11.2 GUIDELINES FOR LESSON OBSERVATION

Excerpted from the Moderator's Guide, TIMSS Videotape Study: http://timss.enc.org/TTMSS/timss/teaching/125445/5445_O5.htm

Following are some suggestions for observing the lesson. Please feel free to move around the room to observe more closely, but please do not interact with the students.

General Suggestions for Viewing the Lesson

Stay focused on the lesson itself: What do you notice? What do you hear? What inferences do you find yourself making and why?

♦ Look for patterns that provides clues to how and what the student/teacher was thinking.

♦ What do you think is the teacher's goal? What does she/he seem to want students to learn? What do you think they are learning?

♦ What does the teacher do? Are there key moves or moments in the lesson? Are there crucial missed opportunities?

♦ Why do you see this lesson in this way? What does this tell you about what is important to you? Look for patterns in your thinking.

♦ What questions about teaching and learning did observing the lesson raise for you?

♦ Are there things you would like to try in your classroom as a result of viewing the lesson? How would you need to prepare yourself and your students to try these things?

Specific Focus

1. Mathematics instruction.

   The first, and perhaps most important, focus area may be the teaching methods used and the learning that students experience.
   Questions About Mathematics Instruction

   • What is the mathematics of the lesson?

   • What seems to be the teacher's mathematical goal?

   • How does the lesson flow?

   • Are there logical connections between the parts of the lesson?

2. Communication between teacher and students.

In analyzing the communication, you may find it helpful to look at the roles of the teacher and students with regard to the mathematics discussed in each lesson.

Again, the point is not to focus on what participants might judge as good or bad, but on what can be inferred about student learning from specific evidence in the lesson.

Questions About Communications between Teacher and Students.

- What does the teacher do to orchestrate discussion in the lesson? What are the questions posed to students? When and how are they posed?

- How do the questions elicit mathematical thinking among the students?

- What does the teacher do to use students' ideas in the discussion? Are most students involved? How are students' ideas used?

- What decisions does the teacher appear to make in regard to students' ideas or discussion? Here, you can probe for more detail by asking:

- Do there appear to be ideas that the teacher is pursuing? Are there times when the teacher decides to provide more information, clarify an issue, model a strategy, or let student's struggle? What do you think that says about the teacher's goal?

- What do the students do in the lesson discussion? What do their verbal and nonverbal communication suggest about their mathematical understanding?

ENGAGING STUDENTS

Here's the setting: Lead teacher Cynthia Sanchez is at the front of the room, her team members, Fran, Magnolia, and Nick, are standing on the side. Twenty-five fifth graders are sitting in groups of four, and, cramped in the back of the room, are 25 strangers like me, eagerly waiting to see action. *Yikes!!! Teachers and students, how can you teach or learn amongst all of this disruption? Good luck Cynthia—you'll need it!* Cynthia starts the lesson by asking students questions on the parts of a circle: "What did we learn on Friday? What did we learn about the circumference of a circle? What do you mean by diameter?" As students respond, she notes the responses on the board.

It's interesting to observe that, throughout the lesson, spurts of Spanish dialogues are interspersed between students' and Cynthia's interactions because not all of the students speak English fluently. For example, when she asks, "What is meant by diameter?" Raphael responds in Spanish. Cynthia follows up with further questions to him in Spanish, and then proceeds to translate comments and responses to English for the rest of the class. *Gosh. This is constant. Yet the lesson flows and students just seem to accept this interaction as quite normal.*

Cynthia tells the students that they will learn a bit more about the circumference of circle in today's lesson. She shows a homemade video of her getting on a Ferris wheel. In the video she tells the students: "Hi, kids. I have a problem I'd like you to help me solve. I am getting on this Ferris wheel, and I will travel all round and back again. Can you help me determine two things: First, how far will I have traveled? Second, how many diameters of the wheel can go into the circumference of this wheel? The diameter of the Ferris wheel is 100 feet."

She waves to the students, gets on the wheel, gets off at the end and asks the students for their help again. Students look attentively at the video as they smile and one says "Hey, that's cool Ms. Sanchez!" *Neat. That has grabbed students' attention. This is not a textbook picture of the wheel, but teachers who have taken the time to make it real.*

Cynthia restates the problem and has students brainstorm how they might solve it. She lists some of their suggestions on the board. The teachers distribute a handout of a large circle with a diameter having a small picture of Cynthia in a box at the bottom of the wheel. Cynthia reminds students that they have a box of materials on their desk to help them with the work. At this point, teachers and guests are invited to move around to listen to students' group processes. *Had all of this instruction taken only 5 minutes?* Students open their boxes and look at the materials: strings, markers, scissors, tape, and calculators. Some students raise their hands for further clarifications. Nick says to one group, "What are we asking you to do? You want to know how far Ms. Sanchez traveled. What do you have in your box that might help?"

Sitting close to where I am is a group of students. I listen carefully and feverishly jot down their comments and then circulate to capture other groups' processes:

Group 1

Santos: How we gonna do this?

Marie: I don't know. I'm stuck. Maybe we should wrap the string around the circle.

Joe: But what would that give you? I think the answer is 200 because, look: The diameter is 100 and it cuts the circle in two equal parts so that this side is 100 (Joe uses his hand to cover ½ of the area of the circle) and this side is 100.

Santos: OK. Let's say 200.

Group 2

Bonita: Let's split the circle in to four parts with the diameter—like a pizza. Now since each radius is 50, the circumference has to be 50 added up four times...or 200.

Angelo: No. Let's split it into eight parts, like a real pizza, and then add them up and get...400.

John: Yeah.

Group 3

Roberto: Let's wrap the string around the circle.

James: But it doesn't have numbers. It's not a ruler. How is that gonna help?

Roberto: Right. I don't know.

Anna: Let's use the string to measure the diameter. We know that's 100.

Roberto: Yeah. Then we keep up with the 100 and see how many times that goes in the circle.

Group 4

Kayla: Let's take the string and measure it around the circle.

John: No. Let's see how long the diameter is first, then do that.

Jenny: Mark it off with this red marker. OK. I'll count. Now, that's one, two, three, and a little piece left over.

It's time now to call the students together to share results. Cynthia asks a member from Groups 2, 3, and 4 to report. From Group 2, Bonita explains how her group decided to use the diameter to cut the circle in to equal parts to get 400 as an answer. Are there any questions?" asks Cynthia. Raphael asks (in Spanish), "Why did you use only eight? What would happen if you cut it up into more parts? Then your answer keeps changing." Bonita replies, "My group decided to use eight." "How many of you agree with this method?" asks Cynthia. No hands go up. Cynthia calls on Roberto from Group 3 to report: "We figured out that since our string doesn't have numbers like a ruler, we measured the diameter to get 100 and cut the string. We measured around the circle and found that it went in three times so we got 300." "We did almost the same thing!" says, Jefferson, excitedly, from Group 4. "Come up and explain." Cynthia says. He responds, "We measured the diameter from top to bottom and that was 100 feet. Then we used the string to go around the circle, and the answer is 300. But we had a little piece left over that we think might be about 21 so we put 321."

Cynthia asks the class: "Well what do you think?" Anna from Group 3 says, "There is no piece left over, so you get 300." "So you got 300," says Cynthia, "If the diameter is 100 and you divide 300 by 100, what would you get?" Anna says, "three." "Hmm," Cynthia continues, "But is it exactly three or three and just a little bit? Raise your hands if you think it's exactly three." While students ponder on what to think, Cynthia tapes a poster of a huge circle on the board and shows the student a long string of four colored ribbons.

> Cynthia: Let's try to answer this question now. First, tell me what you notice about this string of ribbons.
>
> Tania: It has four colors.
>
> Cynthia: Ok. What I did was to measure the diameter and used a different color to represent its length each time. So, this yellow ribbon is as long as the diameter and so it this red, blue, and green. Now we will see how many times I can wrap the string around the circumference of the circle. As I go around, I will tape it so that it stays in place, and then count the different colored ribbons. Amy, count for us, please.
>
> Amy: OK, that's one…two…three…and there *is* some left over!
>
> Cynthia: Yes. Can you all see that I had to use a little bit of the green ribbon to keep going? So. whenever you divide the circumference by the diameter, you get three diameters and a little bit left over. Remember that Ferris wheel from the video? would this be true for it, too? Do you think this relationship is true for *all* circles? Let's try to find out.

What are members of Group 1 thinking? Do they now see where they were wrong? I don't think so. The activity and demonstration were excellent for developing the concept, but I can see from their dazed facial expressions that all of this important discussion is probably having little impact on their understanding. I wish they had been asked to report, too. That might have caused them to reflect further or to pay closer attention or maybe to ask a clarification question.

For the next activity, the teachers give each group of students a set of four circles with different diameters labeled, and Cynthia says, "You are to compute C divided by D for each circle and note your results on the worksheet with the four columns named diameter, circumference, and C/D. As you can see, the diameters range from 2 to 25 inches. *Group 1 is definitely in trouble.*

I look at Group 1 students. Not much activity going on. They got circles labeled with diameters of 2, 6, 12, and 15 inches. One student takes a ruler to measure the diameter of a circle that is already given as 12 inches. Another takes a string to measure the diameter of a circle already given as 2 inches. They measure, then look bewildered, and finally, stop work. The other two members look on. *They really lack basic understanding of the definitions and concepts but if I can get them to review what Cynthia did with the ribbons, maybe that might give them a clue about this activity. Let me give it a shot.* As the audience and teachers circulate to observe the groups, I approach Marie and whisper, "Marie, remember how Ms. Cynthia measured the distance around the circumference of that big circle on the board? What...." Before I could pursue my line of questioning, Patsy Wang-Iverson pulls me over and reminds me to not interfere with the group process. *"Oh No!" the child in me wails, "I've been caught violating the cardinal rule!" This is no joke, readers. I really felt like a kid caught with her hand in the cookie jar.* "Patsy," I say in a whisper, trying to redeem myself, " I was just going to ask some probing questions on the first problem to get them moving. They don't understand the first problem so they are now stuck on the second." "No, don't do that," she whispers, "They may be overwhelmed from all the activity today, but let them try to work it out together. Go over to Group 2, and look at how the students there are working." This is the group that had used a similar parts-of-a-pizza approach on the first activity. Members seem to now understand the process and are actively working and enjoying each other's help. Rather than each student working independently on one circle, they work jointly on one circle at a time: One student measures, another helps tape the string around the circumference, and the third records the solution and uses a calculator for the computation—really nice. I refrain from making any complimentary comments to Group 3 and sit quietly to observe Group 1 again. Patsy poses a few questions and tells the students to work together on the problem. I am hoping to see something occur that will make that happen, but to no avail. They just give up. *I'll need thicker skin for this observer role next time. How can I see this happen and not try to make a difference? I should have called one of the teachers and made them more aware of the problem...but no...no... my role here is to just observe.*

The words, "OK, class," said by Cynthia, take me out of the reflective mode and back into the classroom. She continues, "Let's record your data and then pay particular attention to the *C/D* column. One person from each group will call out the measures. As students do so, Cynthia records them on the board. Because another group also has Group 1's diameters and reports the results, Group 1 does not have to do so. The following dialogue below shows how worthwhile tasks together with skillful questioning techniques can guide students to important discoveries:

Cynthia: Look at the *C* divided by *D* column. What do you notice about all the numbers in the column?"

Eve: Every number has a decimal point and every number has three in it."

Cynthia: Good. What does the three mean?

Joe: There are three diameters in the circumference.

Cynthia: What about the decimal part? What does that tell us?

Anthony: It doesn't go in evenly. There is some left over.

Cynthia: Great! We discovered that, in the Ferris Wheel circle, the circumference was about three times bigger than the diameter or, that about three diameters go into the circumference. Do you think that is true for all circles?

James: That seems to be true.

Cynthia:The Egyptians discovered this and used it a long time ago. It tells us that C divided by D is always the same number. That little piece left over is approximately point 14 or 14 hundredths. What you have all discovered is an important relationship between the diameter of circle and its circumference. This number is actually close to 3.14. Does any one know what we call this number?

Roberto: Pi.

Cynthia: Yes, pi. Now, who can tell me what we learned today?

Amanda: We learned how to compute C over D.

Anna: We learned how to measure without a ruler.

Cynthia: Having gone through the chart of circles with different diameters, do you think you can calculate C divided by D for a circle of diameter 25 inches? Here is a homework sheet to complete."

The bell rings, students leave, and guests proceed to the lunchroom for the colloquium portion. *Why did I have such hard time being a passive observer? Do I tend to help students too quickly and thereby cheat them of the important experience of having to struggle to solve some problems? I'll be mindful of that next time.... No...better yet, I'll get a colleague to work with me!*

COLLOQUIUM

As we move chairs to form a large circle, I notice that some Japanese teachers and Mr. Tanaka, who is the principal of the Japanese School, are present. As is customary in a lesson study colloquium, the lead teacher reflects first and then the other team members offer their comments. Questions and comments from guests follow next, and then a final commentary of the lesson by a designated person, in this case, Mr. Tanaka.

Cynthia: I think I was very nervous for the first part of the lesson. That is why I spoke very fast—too fast sometimes. But I was very happy to see that students finally got the idea of the circumference being three times the diameter and a little bit. I also thought it was very nice how students worked together to come to an answer as a team. I was pleased when a student said that he learned how to measure without a ruler. In the future, I do think that they will remember what 3.14 is and where it came from. The manipulatives were very useful.

Nick (a team member): Cynthia did a wonderful job! I know how hard it is to teach in front of so many people. I think she really used the manipulatives well and stressed the important concepts.

This is the final teaching of this lesson for us. We first tried it using a carousel to pose the first activity but it didn't work well because students couldn't see the entire circumference. The Ferris wheel is a good improvement. I find this process of collaboration so worthwhile and enriching to me professionally. I teach 7th grade, and one of the things that I had to learn was how to talk so a 5th grader could understand. I also learned how important it is to give groups of students one big worksheet or material rather than smaller, individual ones. This forces students to work together because they have to help each other.

Fran (team member and vice principal): I taught this lesson the very first time and find Cynthia's version different in so many positive ways. It was good to make our own videos serve as motivating factors for the students. It really worked well. This collaboration is a phenomenal experience! We all teach the students and they see us working hard to improve our lessons for them. This suggests to them that they too have to work hard. I strongly recommend that you try this approach in your school and practice it across grade levels.

Magnolia (team member): What this process does for us is to take us out of our egocentric world to take a close look at what students are learning. As team observers, we also look closely at the teaching to give advice while it is in progress. We gave advice to Cynthia on which students to call and in what order, so that a variety of approaches could be seen or discussed.

Nick: We would now like to invite your comments or questions on the lesson. We welcome comments on any aspects of the lesson: the use of the manipulatives, wording, theme, sequencing, or any questions you might have that may help us improve on it.

Yvelyne (the author):   I have seen the use of 3D circles with different diameters to help student discover pi, but this lesson's use of the ribbons was an excellent visual. I was also impress with Cynthia's persistence in asking students questions to elicit the main ideas of the lesson. She did a really nice job.

Observer: The ferris wheel video was a great idea. I also liked the model using the cardboard circles and the ribbon.

Yvelyne: I do have a question about timing. I see that each segment of the lesson has a time limit. Is this the reason why every group did not get an invitation to report on the results of the first activity?

Nick: Yes. The first time we did the lesson, we called on every group and found that it took too much time. This time we looked at each group's work and chose one group that had a really wrong answer (Group 2), one not so wrong (Group 3), and the last one right (Group 4).

Yvelyne: But, as a result, students in one of the group's not reporting were not challenged to think through their work and later had no idea on how to complete the next problem.

Cynthia: We don't expect every student to get every lesson, every time. As I teach, I try to get a feel for how the lesson is going to decide on whether I should move on the next day.

Nick: I know which group you are talking about, Yvelyne. You are talking about Marie's group. We anticipated their approach, which was the same as Group 3's. We hoped that reviewing Group 3's work with 8 diameters would help Group 1 see why working with one diameter was also wrong. But don't worry. Students will get to see the concepts presented again in other lesson so that they will get opportunities to think about them again.

Observer: Does a timed sequence determine whether or not you follow up on those teachable moments? For example, when Raphael asks students in Group 3 why they didn't use more diameters in their argument, there was no further discussion on it.

Nick: Yes. I know what you mean. We could have pursued it and talked about what an increasing number of diameters would look like on the resulting inscribed polygon and then maybe get to the point of showing a circle as a polygon with an infinite number of sides. But that would have taken off us off the major goal and we might not have had the time to finish the lesson.

Magnolia: We have to be careful sometimes because so many different lines of interesting questions can come up but we have to focus on our major objective —getting students to learn the concepts in the lesson at hand through the interesting questions related to the objective.

Yvelyne: I was aware of how little you had to "manage" the students. They knew the procedures and what was expected, and they pretty much stayed focused on the task. You merely managed who did what next. This was nice flow of work that revealed lots of good pre-planning and thought. You also always had a question ready for students to consider supporting or refuting. But tell me about Raphael and the Spanish dialogues you have with him and other Spanish-speaking students in the class.

Cynthia: With so many Spanish-speaking kids here, one of my major goals is to make sure that they are not afraid to ask questions. I just automatically flow from one language to the other in a casual way so that it doesn't call attention to itself. Thus we have rich dialogues in both languages.

Observer: I noticed that one group's basket of materials didn't have enough strings to go around so you might consider a ball of string next time.

Observer: At the end of the first activity, Cynthia quickly told students that C divided by D would always give 3 plus a bit rather than get students to think it through. I don't think some students had time to process this most important result.

Cynthia: Yes. You are right. I was so nervous that I did rush through some things. I think that the next time we do this lesson, we will consider spending more time to give students a firm understanding of the first activity's concepts. Maybe the first activity should be the entire lesson?

Observer: Or you might consider reducing the number of circles in the second activity from 12 to 6, thus allowing more time for the first part.

Fran: But with fewer entries in the chart, students' measurement mistakes will likely make it difficult to see the pattern.

Observer: In the lesson, you asked students the key question: Will C divided by D be the same constant for all circles? I wonder how we could structure this lesson so that this question comes from students. Then they will do the investigations to answer their own questions rather than the teacher's. Maybe if they do the chart first and see the pattern, this might motivate that question. But then we'd have to think of a motivating reason to do the chart, too.

Observer: How many lessons have you written after three years of work?

Magnolia: We have completed curricula for grades 7 and 8. By next year we hope to finish a new 6th grade curriculum.

Nick: Colleagues, we will certainly give your comments and ideas a lot more thought. At this point, I would like to thank everyone for sharing observations and ideas with us. We have made notes of them for reflection later. It is now time to have a summary of the lesson from Mr. Koichi Tanaka, Principal of the Japanese School in Connecticut. Dr. Makoto Yoshida will translate Koichi's comments.

Koichi: The video was very good and served its purpose in motivating students. Giving students the big circular board for the first activity was good but caused difficulty for some students to easily put strings around the circumference. I think 3D objects would have been better because they would have been easier for students to handle. You may want to consider the approach of first giving the students the 3.14 as the answer to the problem and then require that students determine how to get it.

I understand that the teacher was nervous, but still, the lesson was much too rushed, and the speed was too fast. She also needs to give a better summary after the first activity. I don't think students got the major point of that activity. The summary at the end of the lesson, however, was very good and likely to help students gain a deeper understanding of succeeding lesson. I noticed that classroom management was excellent and that students were attentive and interested. This was a good lesson.

DISCUSSION BETWEEN COLLEAGUES

Lynn, could you outline the general agendas for teacher meetings and how you allocate time to have them happen?

We have regular grade level meetings to discuss business or other routine concerns. Sometimes curriculum issues also emerge, but most of those issues are addressed during study group meetings or math meetings. Study group meetings are facilitated by any one of the 16 teacher-volunteers. The meetings are scheduled for two hours per week and may center around any issues or lessons teachers want to discuss. Sometimes we may have a guest speaker from the community, or discuss a book or articles that all the teachers read. The grade level math meetings, which Bill facilitates, are one day per week and last 80 minutes for grades K–4 and 40 minutes for grades 5–8. The focus is on helping all the teachers to develop and implement effective lessons. When a teacher needs help getting an idea through to the kids, the teachers asks Bill to coteach or to observe the lesson and provide feedback.

Now, how do I make this happen? Not with substitutes, because we all want to maintain the continuity and foci of our work with the students. Our school has additional teachers from special programs like English as a Second Language, Special Education, Clinton's program to downsize classes, and teacher assistants for the lower grades. We also get student teachers from William Patterson University to work closely with our students. I partner each of my regular teachers with one of these special program teachers. Partner teachers work closely together, so that the students perceive both as regular teachers. Thus, on meeting days, regular teachers leave their students with the special program's teacher, who continues with the lesson. There are times when I may have to call on a guidance counselor or the vice principal to help, but this system generally works fine.

Do you envision the entire school using a study group process if it continues to help students?

This would be a marvelous goal to attain. However, we know that there are many cultural barriers to overcome, such as teachers' tendency to work in isolation, or the tendency of high-stakes testing to motivate teachers to cover, rather than uncover, important concepts.

Do the teachers meet regularly with the other collaborators?

No, but we do see Patsy about once a week. She just pops into a class to observe a lesson or a lesson study. She then follows with immediate feedback through emails.

Who provides funding for the teachers to participate in professional development?

Patsy, through the Research for Better Schools, funds the summer. LSRG helps with substitutes. We also tap our district's funds. We truly appreciate the funding opportunities because they served as catalyst for our process and still enhance what we do. However, it is the teachers' energy and motivation for improving teaching and learning that drive our process. We don't depend on the funding as much because of our modified schedules for working during regular school hours.

Please summarize the data on School 2's state pass rate and comment on the effects of high-stakes testing on your programs.

In the course of the past four years, we used at least three different tests; and students have changed, teachers have changed, and teaching methods have changed. Thus, I must state that there are too many variables at play here to make any statement about causality. Using the same type of test, our pass rate was 57% in 1997, and 77% in 1998. Using a harder test, it fell (as was true for other districts and the state) to 46% in 1999 and 40% in 2000. We are still below the state pass rate, but we are scoring at or above the district's rate.

I must stress that we are not letting the pressure to raise test scores adversely affect what is best for our students. Our major goal is not to improve test scores but to improve the learning of our students. We hope and trust that ultimately what we are doing will positively effect scores; however, we know and accept the fact that this is not necessarily an immediate outcome. We believe that real lasting change requires time for teachers to reflect deeply on how students learn, as well as on the curriculum and how it is to be taught. Ours is thus a long-term process of improvement, and definitely not a quick fix.

Teachers: Are more students doing homework?

Nick: Yes but a major reason for this is that we don't assign as much homework as we used to—just a few problems for practice on major ideas. Students are thus enticed to give it a try.

Has the study group process impacted your math content knowledge?

Nick: Definitely. None of us has a certification in mathematics even though some of have enough content knowledge to do so. Those of us who would qualify forgo that process because automatic placement in high schools where math teachers are in dire need, is almost a given. We thus have a good mix of colleagues willing and eager to help others whose content knowledge maybe weaker.

Where does the work grow from here?

Bill: Well we will continue to improve our 7th and 8th grade lessons through lesson study, and we will try to maintain our partnerships.

Extending Japanese lessons to the lower grades may not be as daunting because Singapore uses the lesson study process, and their texts for grades 1–8 are in English. We have begun to use these texts in the lower grades and are having positive experiences.

Any further comments?

Lynn: Professional development is a must for improving education of our students. Yet we seem to change everything except for the core. Lesson study focuses on that core.

Patsy: In a 1998 focus group of students, which I conducted, a student said, "Before the book did the thinking. Now, we do the thinking."

Frank: Yvelyne, you should know that you are not the only American observer that had a difficult time just observing the lesson. I noticed the same thing among most of the other American observers during the conference. They wanted to serve as tutors, while the Japanese teachers made notes about the lesson. I think that is the key difference in the training for the lesson study. Even though they were given guidelines, the American teachers did not know what to do as a lesson observer, as they seldom do such, while the Japanese teachers were in fact trained to be observers.

Yvelyne: My jaws dropped when I read your comments, Frank. *Trained* to be observers? Who would think to do that? Is it part of our cultural script that we think we *know* we can do that easily? I don't think that I would have thought training was necessary had I not experienced the pangs of passive observation. I can see how right you are. Thanks- I better understand my reaction and can easily connect it to problems my students encounter when conducting clinical interviews to assess a child's understanding. In spite of "how-to" instructions, they often slip from an assessment mode, to one focused on teaching.

COMMENTARY

Given that Public School 2 teachers said the Japanese method of teaching was new to them, I think the method has possibilities for reforming teaching in the U.S. Indeed, Stigler and Hiebert (1999) write, "In our view, lesson study is not the kind of process in which teachers must first develop a list of capabilities and then begin to design improved lessons. Lesson study is, in fact, the ideal context in which teachers develop deeper and broader capabilities" (152).

As I reflect on Cynthia's lesson, I ask the question: How different is her lesson from those profiled in this book? I wrote the lesson plans (overviews), and a look at both plans shows definite differences in the depth of detail provided. Different elements in the lesson study plan include minutes allocated to each section, students' anticipated responses, and an evaluation column. Evaluative, focused questions value both process and affective results in the assessment of students' learning, while providing acute reminders to keep a focus on students' learning, and also serve to guide the team teachers' later reflections and comments about the lesson. For example, "Are students using their prior knowledge to solve problems? Are they anxious to begin working on the problem?" Unlike the plans in this book, which I now see as static, the lesson study plan has a dynamic quality, which is projected by its details for monitoring teacher and student interactions throughout the development of the lesson. For example, anticipating students' responses shows how this plan "moves" towards keeping students at the center of the lesson. What strike me, too, is the effective use of columns to organize the major pieces so that a glance quickly shows what teacher and students are doing at any given segment, as well as whether students are behaving in a way that hints at an understanding of that segment. Thus, the lesson study plan also *visually* seamlessly integrates teaching, student learning, and assessment, while it prepares to provide answers to elements that may hinder or enhance learning. Wow! What caring teacher's teaching would not show some noticeable improvement from carrying out such a plan? What caring teacher's teaching would not show *remarkable* improvement from participation in the development and implementation of such a plan?

As I focus on the classroom environment, however, I surmise that there is little or no difference in the *learner desired outcome* and *actual classroom teaching* of the Japanese method when compared to what I will call the "NCTM method" depicted in this book. Indeed, in his examination of five classroom situations in Japanese elementary schools, Sawada (1997) concluded that the situations demonstrated the Japanese teachers' use of the concepts in the *Standards* documents. What School 2 teachers show is that with appropriate support and collaboration, American teachers can move closer to applying the NCTM method and can make a positive difference in students' learning and attitude towards mathematics. The *Principles and Standards* (19) make explicit reference to the value of such collaboration:

> Reflection and analysis are often individual activities, but they can be greatly enhanced by teaming with an experienced and respected colleague, a new teacher, or a community of teachers. Collaborating with colleagues regularly to observe, analyze, and discuss teaching and students' thinking or to do "lesson study" is a powerful, yet neglected form of professional development in American schools (Stigler and

Hiebert 1999). The work and time of teachers must be structured to allow and support professional development that will benefit them and their students.

In answer to her question, "Does Lesson Study have a future in the United States?," Catherine Lewis (2000) writes,

> [M]y question of whether lesson study has a role to play in the United States is an emphatic "yes," though I think we will need to find the most effective ways to adapt it to our cultural settings. The graveyards of educational reforms are littered with once-promising innovations that were poorly understood, superficially implemented, and consequently pronounced ineffective. If lesson study is to be any different, it will require a deep understanding of what it is and why it has been useful to Japanese teachers, and how it can be adapted to our very different settings" (19).

Efforts to provide part of that deep understanding is being done through research conducted by the Lesson Study and Research Group at Teachers College.

Finally, Hoff (2000) summarizes the impact of the efforts of School 2 staff with its collaborators:

> At School 2, educators are seeing enough progress to keep going. Last year, the 8th graders at the school passed the state math exam at a higher rate than the Paterson average. When the school's students move into high school, they are more likely than students from other schools to enroll in Honors Algebra and Algebra 1. Last year, all of the School 2 graduates who took the honors course passed, as did three-quarters of its graduates in Algebra 1. Says Smith, the Teachers College professor: "This came from a school that was declared one of the four worst in the system, and is now one of the most creative and productive."

As is evident from my report on the lesson study conference, I thoroughly enjoyed it and learned quite a bit. The mere fact that guests were allowed to observe actual classroom, is in itself a great feat because School 2 conducted school *as usual* during that day. I wish to thank all involved, especially the teachers and principal Lynn Liptak.

CONTACT

- School:

  Public School No. Two
  22 Passaic Street
  Paterson, NJ 07501
  Phone: (973) 881-6002 Fax: (973) 238-0936

- Principal: Lynn Liptak: Lliptak@aol.com

- Vice Principal and Team Member: Fran Dransfield

- Middle School Study Group Teachers:

  William Jackson: wcjackson@mindspring.com

  Isabel Lopez: Imahepad@aol.com

  Magnolia Montilla: Briabelle@aol.com

  Beverly Pikema: Bevep@aol.com

  Cynthia Sanchez: Sanchez_cynthia@hotmail.com

  Nick Timpone: Ntimpone@aol.com

- Collaborators:

  Patsy Wang-Iverson: Wang@rbs.org
    She is a senior associate for Research for Better Schools. The Phil-
    adelphia nonprofit organization runs the regional Eisenhower
    Consortium for Mathematics and Science Education.

  Professors at Columbia University Teachers College:

  Columbia University Teachers College
  Box 118, 525 W. 120th St.
  New York, NY 10027
  Phone: (212) 678-4166

  Frank Smith: F1s12@columbia.edu

  Clea Fernandez and Makoto Yoshida: cf170@columbia.edu
    They direct the Lesson Study Research Group at Columbia.
    Yoshida's work (Yoshida 1999b) highlights lesson study devel-
    opment in Japanese schools through case studies.

  Koichi Tanaka, principal of Greenwich Japanese School, CT.
    School 2 teachers get help in running lesson studies and receive
    advice from the Japanese educators.

The Greenwich Japanese School
The Japanese Educational Institute of New York
15 Ridgeway
Greenwich, CT 06831
Phone: (203) 629-9039 Fax: (203) 629-9660
E-mail: Gwjs@earthlink.net

FIGURE 11.3 RESOURCES AND REFERENCES

Dear Readers,

If you have some questions left unanswered, try the contact information, as well as the references below. Also, for a well-written description of the 7th grade AMTNJ lesson on the Fibonnaci sequence, read Hoff's article.

Resources and References on
School 2 and Lesson Study

Coeyman, M. (2000a). Changing America's path to perform. *Christian Science Monitor*, May 30, 2000.

—(2000b). U.S. school, Japanese methods. *Christian Science Monitor*, May 23, 2000. http://www.csmonitor.com/sections/learning/mathmelt/p-4story 052300.html

—(2000c). How Japanese students learn math: Teachers get good results with group work, problem-solving approaches. *Christian Science Monitor*, May 23, 2000. http://www.csmonitor.com/sections/learning/mathmelt/p-5story 052300.html

Cooper, R. (1999). A call to focus on techniques of teaching. *Los Angeles Times*, September 22.

Hoff, D. (2000). A teaching style that adds up. *Education Week*, February 23. http://www.edweek.org/ew/ewstory.cfm?slug=24timss.h19

Lenz, B., K., Deshler, D.D., & Schumaker, J. B. (1990). The development and validation of planning routines to enhance delivery of content to students with handicaps in general education settings (Progress report). Lawrence, KS: University of Kansas Center for Research on Learning.

Lewis, C. (2000). Lesson study: the core of japanese development. Invited address to the Special group on Research in Mathematics Education American Educational Research Association Meetings, New Orleans, April 2000, session 47.09.

— (1995) *Educating Hearts and Minds: Reflections on Japanese Pre-School and Elementary Education.* New York: Cambridge University Press.

—Tapes featuring Japanese lesson study group teachers. Contact Shelley Friedkin, phone (510) 430-3350.

Ma, L. (1999). *Knowing And Teaching Elementary Mathematics: Teachers' Understanding of Fundamental Mathematics In China and the U.S.* Hillsdale, NJ: Lawrence Erlbaum.

Schmidt, W.H., McKnight, C.C., & Raizen, S.A. (1997). *A Splintered Vision: An investigation of U.S. Science and Mathematics Education.* Boston: Kluwer Academic Publishers.

Stigler, J., & Hiebert, J. (1999). Understanding and improving mathematics Instruction: An overview of the TIMSS video study. *Phi Delta Kappan* 79:1, 14-21.

—(1997). *The Teaching Gap.* New York: Free Press

Yoshida, M. (1999a). Lesson study: An ethnographic investigation of school-based teacher development in Japan. Doctoral dissertation, University of Chicago.

Yoshida, M. (1999b). Lesson Study in elementary school mathematics in Japan: A case study. Paper presented at the American Educational Research Association Annual Meeting, April 1999, Montreal, Canada.

12

THE NCTM VISION: MAKING IT HAPPEN

IMPACTING PRESERVICE TEACHERS: REFLECTING ON NEEDS

I do not know why, but I feel as though I have spent the last year living and breathing math, when in actuality, it has really been three short weeks. Every time I tell someone that I am taking a mathematics course over the summer, I get the usual "ugh, you poor thing!" or "what a horrible way to begin your summer." I admit—when the semester first began, I was an active member in that self-pity club. Now that the course has almost come to an end, I wonder where all the time has gone. I cannot believe what I am about to admit, but I actually enjoyed the past three weeks in this class.

This course has not only reaffirmed my teaching philosophy, but it has added to its truth and strengthened it. My thoughts have not changed one bit. Moving away from a traditional learning style I now realize that actions do speak louder than words. Children need to touch, see, and manipulate information in a way that the information is presented as a challenge. Just like video games, children want to master challenges and move on to the next challenge, or level, that awaits them. Effective teachers have the ability to flip on a certain switch inside a child's head, advancing them to the next level.

It seems that I have been able to connect so many things that were scattered around in my mind through this class. I could better understand my math anxiety, and realize that I was not alone. I could connect my past failures with reasons and better myself through this. I could see how manipulatives could open my eyes to a concept that I had always found so difficult.

I cannot believe how much I have learned these last three weeks! Aside from working with students my own age and kids from K–1, 2–3, and 4–5, I have also taken my knowledge and applied it to an outside experience: I actually used pattern blocks to teach my neighbor fractions! Rarely have I ever extended schoolwork to other aspects of my life, and I had a wonderful time doing it.

I feel so fortunate to have met and befriended this special group of classmates. They have shared so many things with me, and many ideas that I will definitely use in my future classroom—from fun and successful math lessons to classroom management tips. I have also been able to share with classmates things that have and have not worked for me. I probably learned the most from these class group discussions.

I am no longer afraid of math. I feel so much more comfortable with things that I was petrified of before this summer semester. Of course, I still have a lot of questions and concerns, for example, will I ever be able to effectively teach certain math concepts that I am uneasy with myself? Only time will tell, but with the foundations like this brief, but thorough, math course, I think that I will be OK.

Brittany,
Preservice K-8 teacher

Brittany's reflection on her experiences in a three-week mathematics methods (not math) course shows how big an impact a well-structured math methods course can have on helping a beginning teacher attain an appreciation for the important elements of the NCTM vision. Based on Brittany's comments, we can surmise that the course addressed aspects of NCTM's vision for a reform mathematics program in a number of ways: The curriculum was rich enough for Brittany to consistently refer to it as a "math course"; students were taught using a variety of approaches aimed at enhancing their conceptual understanding of important concepts—so much so that Brittany felt confident enough to immediately teach others what she had learned; and students learned in an environment that invited them to form a community of learners interested in helping its members share and resolve concerns. This environment helped Brittany to let go of her anxiety and to accept the fact that she has resources to help her when she needs it. For a view of the syllabus for the method's course, go to http://www.tec.uno.edu course syllabus. Under Summer 2000, find EDCI 3140.

Brittany's fear of inadequate content knowledge, however, is of concern not only to her but also to the larger mathematics community. In her book, Liping Ma (1999) follows up on this research and the TIMSS study by investigating mathematical understanding of U.S. and Chinese elementary teachers. Her interviews of the teachers show that, in general, the Chinese teachers have a better understanding of the mathematics they teach than U.S. teachers do. Ma writes that the Chinese teachers have a "profound understanding of fundamental mathematics" (PUFM), which she describes as being "more than a sound conceptual understanding of elementary mathematics—it is the awareness of the conceptual structure and basic attitudes of mathematics inherent in elementary mathematics and the ability to provide a foundation for that conceptual structure and instill those basic attitudes in students" (124).

How can teachers help students like Brittany gain a deep understanding of mathematics? In Chapter 1 of this book, and throughout the profiles, we find answers that are within reach of every K-college mathematics teacher.

REVISITING QUESTIONS

In Chapter 1, my preservice teacher Angeline described her daughter Georgine and posed an interesting question in her reflection: "Georgine's passion lies in social studies and literature. She is not a math-brained child, I guess. Are these children born, not made, that way?" My comments to her expressed these thoughts: At the heart of teachers' attempts to reach all students is the belief that all students can succeed in mathematics. The task is to find in what ways our students are smart, and how our teaching can be adjusted to extend their strengths to the area of mathematics. For Britanny, it was the opportunity to play with manipulatives that allowed her to bridge the algorithms she learned to the meaning underlying the rules.

CAN IT BE TRUE THAT ALL STUDENTS CAN DO MATHEMATICS?

My comments to Angeline motivated me to review the literature to support my ideas on this question, which is also the title of an insightful article by McGhan (1995). One reason McGhan answers "yes" is his conviction that with effort, teachers can make a difference. However, he writes:

> Many will still wonder how this viewpoint squares with reality. They will think something like, "you can't make a cow jump over the Moon just because you want her to." And we can't, because it is certainly true that there are some people who are profoundly mentally handicapped. Estimates usually peg this proportion at about 2% (this percent can actually increase beyond those with genetic disorders since prenatal conditions [e.g., fetal alcohol syndrome, poor nutrition] and conditions in life [e.g., accidents, abuse] may cause profound mental

deficiencies.) At the other end of the human spectrum, there are people like Einstein....So if necessary, we might say there is another 2% or less at the other end of the intellectual spectrum who seem profoundly different from most people. That leaves 96% or more of us who might be just about the same....Whether this model of human capability (it's similar to a model Usiskin [1994] calls the squeezed normal distribution) is accurate, is unknown. Here again we can make existential choice: I choose to believe that there are a few mentally handicapped people in the world, and a few truly gifted people, but that most of us are very much alike in intellectual capacity.

My colleagues in the Department of Special Education and Habilitative Services would add to McGhan's existential choice. They reason as follows. From a physical perspective, the brains of some students at the bottom portion of the 96 percent may appear to be severely damaged. However, the extent of cognitive debilitative damage cannot be fully measured by observable behaviors and is therefore difficult to diagnose. There is always the possibility that the assessment tool that was used was inappropriate for that particular individual. They also note that common questions in their field about such students are, "When did the student learn that?" and "Did you know that the student could do that?" Because learning occurs over a period a time, the belief that all students can learn is key to the ability to provide all students with the opportunity to reach their potential in any subject. Finally, a flexible curriculum and an assessment process that supports all the ways people can learn and allows for individual differences is crucial (Burrell et al. 1998). How much their reasoning calls to mind the *Standards!* Their recommendations are all represented in NCTM's *Standards* documents.

CAN ALL STUDENTS DO ALGEBRA?

After the preceding section, it might seem irrelevant for me to ask this question of any topic in mathematics. However, the caution by special educators to look at curriculum when debating questions about all students is particularly relevant to algebra because many states now mandate it for all students at the 8th grade level. Indeed, prominent mathematics educators voice concerns that lack of attention to the algebra curriculum may result in failure for a large number of students. Silver (1997) writes,

Mandates can create a situation in which students have greater access to an algebra course but one in which they are likely to fail....If the goal for students is their attainment of fluency with algebraic thinking, then mandating algebra enrollment for all students has another likely limitation—the course itself. The increased pressure to offer algebra to a large number of students is likely to result in more students taking a traditional algebra course...(which) does not ensure that they will gain access to the important ideas that lie behind the seemingly endless list of procedures that are often taught as if they were disconnected from meaning. (206)

Silver's statement clearly implies that extending the traditional algebra course over several semesters for students who may need additional time to understand ideas will not resolve the problems. Mathematics educator Cathy Seeley emphasized this point in the NCTM panel teleconference during the 1998 Annual Meeting in Washington, DC (*Algebra for All*, NCTM/PBS Mathline, April 1998, phone: (202) 842-0552):

> The trend in algebra over the years has been to push it down, slow it up, speed it up, accelerate it, remediate it, enrich it, and generally track the heck out of it. What I have discovered is that if you take a year of traditional, abstract, boring algebra and spread it over two years, you get two years of slow, traditional, boring, abstract algebra.

Thus, the traditional algebra curriculum is not what prominent mathematics educators recommend for all students. Silver views NCTM's recommendation that algebraic ideas be integrated throughout the K-8 mathematics curriculum as a key element that will help students achieve algebraic competence.

In the interim, what about algebra for the students we have *today*? Some have not been through a reformed curriculum or even mastered the traditional basic skills. How can we expect them to succeed in algebra? One of my preservice teachers, Sergio, was confronted with this problem while he tutored Marie, an 8th grader who could do the steps for solving $6x + 1 = 73 - 3x$ up to $9x = 72$. She knew to solve for x by dividing by 9, but she could do so only with a calculator. Although Sergio's inclination was to insist that she put the calculator away, I advised him to let her continue to use it, for two reasons. First, Marie will not be able to complete all of her homework during tutoring if she has to reconstruct multiplication and division for every problem. This implies that she will not be prepared for the next class, which, in turn, may delay her understanding of that day's algebra lesson. Second, the fact that she encounters the basics again in algebra is another opportunity for her to master them, but from a different perceptive.

It was Sergio's task, and his challenge, to find ways to help her learn and practice the basics she had missed while at the same time inviting her to continue her explorations of an abstract world that some believe is accessible only to those few people who have a "math brain." Marie's challenge, which she made a priority during her study of algebra, was to learn strategies to memorize the traditional basic facts so that she would not remain dependent on her calculator.

This episode reminds me how fortunate I was that my own high school did not have the two-year prealgebra or general math classes that did little more than rehash traditional middle school mathematics. In middle school, I firmly believed that you don't study for mathematics—you either know it or you don't. I entered high school excelling in every subject but mathematics. I understood, liked, and passed mathematics, but I had not memorized the basic facts. In spite of my low grade, I was placed in the only 9th-grade math course available at the time—algebra. Luckily, my algebra teacher, Mr. Wineberger, said to me one day, "Based on your class work, I'd expect excellent grades from you on exams. What's going on? Do you study for the tests?" That question changed my perspective on the nature of mathematics, motivated me to memorize and study facts for exams, and led me to pursue a major in mathematics and mathematics education.

Two factors encouraged me to continue my study of mathematics—having access to algebra, and having a teacher who believed that I had the potential to excel. Comments from panel members at the 1998 NCTM teleconference suggested that my "luck" must become commonplace if all students are to succeed in algebra. Panel members gave examples that support the view that a lack of basic computational skills should not deny students access to algebra. In one of his overhead slides, mathematics educator Lee Stiff described how students at one school in North Carolina who were given the opportunity to study algebra met the challenge. Contrary to what some might believe, when the number of students taking Algebra 1 increased, the level of student achievement stayed the same. He said, "What was needed was a commitment on the part of the school system to provide all students with access to algebra."

Stiff also showed a slide that stressed the powerful role that teachers play when they decide *who* takes algebra. He listed names of students who showed potential for algebraic reasoning on national percentiles and standardized tests but were *not* recommended for algebra by their teachers. The slide showed the students' scores—but not, he noted, "who these students are, and how these students behave, and where those students come from. Many teachers use *those* criteria to make decisions about whether or not students go on to take algebra: *They misbehave in class; They don't talk like I do; They wear funny clothes and funny hairdos.* I think these are the wrong criteria for deciding whether or not a student takes algebra."

We must all enter into the essential discussion about the specific mathematics content that all students should have, how to create opportunities for all students to make sense of them, and how to assure that all students are encouraged to learn them. Although the first two issues are addressed in *The Principles and Standards for School Mathematics*, the third requires that we reach out to students as Mr. Wineberger reached out to me. That cannot be achieved through standards—only through caring teachers.

CURRICULUM CONCERNS OF A PARENT

The above discussions show how curriculum, assessment, and teachers are all important in determining the number of students who do mathematics. Of the three, I would place the teacher as the most influential factor—it is the teacher who must decide how to shape curriculum or assessment tools to meet the needs of students. In her reflection, Angeline considers the potential impact teachers can have by modifying curriculum to reach all students. As she further reflects on whether her daughter Georgine could develop an understanding of mathematics that extends beyond rote memorization, she writes:

> The reading and class discussion helped me to answer this question. First, I am now aware of the various misconceptions students have about math, and some of the methods for motivating students. Second, we discussed how to utilize the different abilities each student has to help the class as a whole, and also to develop their own understanding of concepts. I know that what would interest Georgine is the history of mathematics and of how mankind first developed a numerical system. This is probably where the teacher can use that part of Georgine's brain to develop the other part and hence create an interest.

It is critical to Georgine, and to Angeline's future students, that Angeline acknowledge that one part of the brain can influence the other. I'd add that if Georgine's teacher allows her to pursue questions in the areas of mathematics that are closest to her interests, Georgine may, from Angeline's perspective, become a "math-brained" person. For that transformation to occur, Georgine's teacher would have to guide her in connecting history to the mathematical content under study. If we believe that mathematics is everywhere around us, then with some effort, we can help students find a mathematical connection to "hook" them into learning.

WHAT MIGHT A TWENTY-FIRST CENTURY SCHOOL LOOK LIKE?

We have looked at the classroom environment from the perspective of teachers who plan and implement lessons that are designed to prepare students to become independent and critical learners. But what about the physical space in which learning might occur in the future? What changes in teaching and learning are necessary when schools are not confined to the traditional classroom? Caine and Caine (1997) describe a learning environment developed by Creative Learning Systems Inc., a high-tech research and manufacturing company based in California (10966 Via Frontera, San Diego, CA 92127-1704; telephone: 1-800-458-2880). Information from the company depicts an environment very different from the traditional classroom:

> A Creative Learning Plaza is a large open environment for up to 150 people, working together for several hours at a stretch. In this environment everyone is a learner and both young people and adults can be facilitators....As you enter the expansive space, video monitors and colorful, glowing light boards flash information about the highlights of the day. Moving into the main area, you can see pathways winding among islands of intense activity. People are busy everywhere, their diverse pursuits spilling over into special enclosures that form the perimeter of the Plaza. Groups of learners are intently building, planning, testing, organizing, debating, conferring and presenting....Learning is project-based rather than subject-based, so session duration is determined by the scope of the project....By having contact with experts and resources anywhere in the world, learners and facilitators have ready access to any relevant knowledge that they require....From digital video to biotechnology, the principles that are introduced, explored, and manipulated in the Plaza are of utmost relevance to our society. Math, science, language, and social studies are integrated into the learning experience in such a way that arbitrary boundaries disappear. Five teachers, serving as facilitators, move around the Plaza to guide and keep groups in touch with one another. A CBS special broadcast described it as a "school for the twenty-first century."

Are we ready for this kind of environment? According to Caine and Caine (1997), "The Creative Learning Plaza will call for what we mean by brain-based teaching. It also matches our view of one way that schools of tomorrow may be organized. We also believe that teachers will not be able to function in this type of environment without a radical view of learning and teaching." An example of such a school is in Alameda Unified School District, Alameda, California. During my conversations with one of the directors of the company, he added that students, too, are not ready for this environment. In fact, he commented that in Alameda County, it can take from four to six months before students are able to function independent of a teacher's constant lead. Once they are comfortable with teachers who behave not only as facilitators but also as learners, students are then able to use the technological tools to solve problems they generate. What role can segments of the community play to make that happen?

WHAT CAN CLASSROOM TEACHERS DO?

As teachers, we may find our task overwhelming. Not only must we change our thinking about the nature of teaching and learning—we must also help students change their thinking as well. At the middle and high school levels, many students are not prepared to learn in the way the teachers in this book teach. The reality, however, is that we can not wait for reform to take place at all levels before we join the movement. We must start with the students we have in our classes today.

The teachers profiled in this book exemplify some of the things we can do. They include becoming a member of NCTM; reading reformed-based articles in journals; participating in conferences; establishing and maintaining contacts with other colleagues; trying, revising, and reflecting on lessons; listening to and assessing students as a basis for informing instruction; and sharing our views with parents. All of these are key actions that will help to reform teaching and improve student learning. Books or readings that detail teachers' accounts of classroom practices and the kinds of problems and issues they face daily are worthwhile reading (Schifter 1996; Schifter and Fosnot 1993). Electronic resources include NCTM's Web site (www.nctm.org). It has information ranging from job listings to hot topics in mathematics education to forums for discussions of important issues in mathematics education to the text of the *Standards*. Finally, we must remember that there is no algorithm for a quick and painless change toward reforming our practices. We are also not expected to do it alone or to do it all every day.

WHAT CAN ADMINISTRATORS DO?

It is important that teachers and administrators work together to devise strategies to move their districts toward implementing reform-based changes. To do so, administrators must be aware of reform issues. In addition to the resources cited above, administrators can find supportive services at the national level. For example, school districts that need help to select or implement curriculum projects may contact the K-12 Mathematics Curriculum Center. The Center offers a variety of products and services to help school districts select and implement curriculum projects funded by the NSF (see the Appendix).

The history of reform in mathematics education shows that teachers are crucial to reform and that, without their support, educational reform fails—no matter how good a curriculum is. Gaining that support requires that teachers be part of the discussion that shapes their professional development program and that they be given the time to reflect and examine their own teaching practice. In addition to specific professional development opportunities, administrators should make it possible for teachers to build time into their schedule to discuss and reflect on teaching during the school day. As a teacher who encourages his preservice teachers to write "scripted lessons," Lee (2000b) writes, "However, unfortunately, once in the classroom, my former preservice teachers frequently stop writing scripted lesson plans in order to cope with a variety of competing concerns such as grading homework, constructing tests, managing instruction, and monitoring students' behavior, along with bus and lunch duty, after school tutoring, and extra-curricular clubs. Somehow we must change the school day so that focused attention to the critical aspects of teaching a lesson can be emphasized" (3). Some schools provide such opportunities for teachers by beginning the school day earlier to allow 90 minutes each week for teachers to meet and plan for instruction. Other schools create a schedule by which all teachers at the same grade level have a common planning period.

Although helping teachers acquire the new strategies, resources, and techniques represent the greater part of the move toward reform, it is not the whole picture. It seems obvious that the process of maintaining and updating the resources that are provided for teachers is crucial; however, it cannot be stressed too much. An email message from an exemplary teacher, who was profiled, shows frustrations that are common occurrences for too many teachers. Below is the teacher's response to my casual question about how the semester was going:

> I am having an awful year! Long story, but my computer science classes are an absolute mess! Unfortunately, the problems are no hardware support, networking bugs, and no administrative support. Therefore, so much of this is out of my hands. If I could quit right now, I would! But, I can't, so it is back to the planning stages as I attempt to figure out how to teach a computer science class without reliable computers. Luckily, I also teach two math classes. We are using Dale Seymour's *Probability Through Data* book at the moment and it is going beautifully! At least I have two hours of some positive feedback—the rest of the day I feel like an idiot!

The good news is that the need to change is recognized and accepted at many levels of the education community where resources and funds are available to support the process of change. Administrators who join forces with teachers, college and university faculty, and others in the community to reallocate or to apply for funds will help solve some of the resource problems in schools and also help to form a more cohesive community.

WHAT CAN TEACHER EDUCATORS DO?

Assisting preservice and in-service teachers to implement the NCTM reform documents should be a major goal of a teacher preparation program in mathematics education. Leadership classes or workshops to prepare teachers to become the leaders that are envisioned in the reform documents are also necessary. In the past, most teachers were handed detailed guidelines to follow that adhered closely to a textbook. Shifting from the mindset of having all the plans "out there" to one where they are *expected* to have input is not automatic or easy. Because they are to facilitate students' learning and engagement in productive planning discussions in their classrooms, teachers need the opportunity to facilitate their own learning as well as the discussions of their peers. Courses that provide time (which may not be available during school hours) for teachers to come together in school teams to vent, discuss, and move toward tackling relevant issues should be offered. Such courses would provide avenues for teachers to construct knowledge that would be based on events from their classroom and related research.

While the above actions will better prepare teachers to teach, more is needed to enhance their mathematical content knowledge. As is true for Brittany, some students completing the requisite contact hours of mathematics courses in the mathematics department do not necessarily enter a methods course with conceptual understanding. Further assistance should extend beyond providing teachers with experiences to apply the standards in the context of their methods course to the harder task of making it possible for teachers to be taught from a standards-based perspective in their mathematics content classes. Gathering faculty from math content and math methods departments who are interested in reform issues and are willing to meet for the purpose of improving teacher education is a good start toward effecting changes in the departments. Such collaboration and can lead to grant applications to help faculty test or implement some of the ideas generated by the group. Inviting teachers and administrators to join discussions at this level is important in efforts to forge a common vision to help guide and connect reform-based changes at all levels. Some of the profiles in this book demonstrate how such a professional community can extend reform into the classroom to positively affect teachers' practices and students' learning.

In Louisiana, the Greater New Orleans Collaborative (GNOC) is an example of a consortium of universities that hosts conferences and workshops to support reform-based instruction in mathematics and science at the college level. Some university faculty are implementing the same process that is recommended to move teachers toward reform: engaging in professional development activities, creating a common planning time to discuss reform issues, and participating in collegial coaching to try new ideas. Support to launch the GNOC originated from the Louisiana Collaborative for Excellence in the Preparation of Teachers, which is funded by NSF and the state of Louisiana.

WHAT CAN WE ALL DO TOGETHER?

Incentives at all levels are necessary to support reform. They can range from a parent who thanks a teacher for taking time to inform the parent of the new curriculum, to a teacher who welcomes a student's insightful comment in the classroom, to an administrator who provides time and compensation for a teacher working to move toward a more reflective practice, to the assessment of a policymaker that a faculty member's work on reform issues *is* an area to be included among productivity accomplishments for promotion and tenure decisions. More global incentives, however, require that all of us have a common vision and a coherent plan for implementation that reduces, if not eliminates, conflicting or contradictory messages to members of our professional community. Having a shared vision of the challenges and successes of teaching and learning is necessary for effective communication amongst us. Books or other readings that detail teachers' accounts of classroom practices and the kinds of problems and issues they face daily are a place to start (Ball, 1998, 1997; Schifter 1996 a,b; Germain-McCarthy, 1999; Lampert and Ball, 1998; Simon et al., 2000). But ongoing communication must also take place to refine and illuminate various perspectives important to promoting coherent policies. An NSF strategy to promote such coherence is modeled in regions of states that have acquired its Systemic Initiatives Grants: Representatives of stakeholders from all levels of the education community come together to discuss and implement actions conducive to reform. Without an influx of dollars to spur such a movement, however, local district leaders may have to start with a smaller piece of the larger puzzle by replicating this process in their own districts.

Consideration of incentives also implies that there is societal appreciation for what we do. Lee Iacocca said:

> In a completely rational society, the best of us would aspire to be teachers and the rest of us would have to settle for less, because passing civilization along from one generation to the next ought to be the highest honor and responsibility anyone could have.

We know that many of our best are not choosing teaching because of its poor status in our society. I believe we can partly reverse this trend by doing a better job of clearly and explicitly promoting the essence of Iacocca's quote. For a start, my new business cards will boldly read "*Teaching* is the highest honor and responsibility anyone could have!" In Checkley and Kelly's (1999) interview of Asa Hillard, Hillard states, "I see people in all kinds of jobs who are unhappy...much of what lawyers do for work is work that they don't *want* to do....We teachers do have incredible working environment—places of learning that are place of joy—when the environment is properly structured. We don't tell people that. We need to tell people how good teaching is. We have—literally—one of the best jobs that anyone could have" (62).

CONCLUSION

The profiles in this book show teachers and students learning new topics through new forms of collaborations and interaction. As they model how to translate the *Standards* documents into real and workable classroom practices, the teachers also model how to engage students at multiple levels of interactions and understanding. Although most of the lessons occur in a regular-sized classroom, the teachers and students participate in a process that promises a smooth transition to a Creative Learning Plaza. Just as important, for those students who may not learn best in such a plaza, the teachers demonstrate a process of teacher-to-student interactions that yields insights into how they might reach such students, and hence, *all* students.

We can be assured that we have taken a giant step toward helping students become self-directed individuals if we can get them to move away from blurting, "You never taught us that!" when they are faced with a novel situation and toward reflecting on the question, "What resources do I have or can I get to approach or solve this problem?" I hope this book helps readers to visualize how the role of classroom leader can promote an environment that supports all students' creativity. I end (begin?) with the last paragraph from Angeline's reflection:

> This week has been one in which I thought about the future of my children, Louisiana's children, and the nation's children, and the power teachers have in steering them in the right direction—given the right tools. I think we are headed in the right direction. We just all need to pull together.

Please feel free to contact me, the profiled teachers, and publisher Robert Sickles to share thoughts, successes, or concerns.

Yvelyne Germain-McCarthy
University of New Orleans
Dept. of Curriculum and Instruction
New Orleans, LA 70148
Phone (504) 280-6533
e-mail: ygermain@uno.edu

Robert Sickles, Publisher
Eye on Education, Inc.
6 Depot Way West
Larchmont, NY 10538
Phone (914) 833-0551
e-mail: sickles@eyeoneducation.com

APPENDIX

NSF CURRICULUM MATHEMATICS PROJECTS

In 1997, Education Development Center, with funding from NSF, established the K-12 Mathematics Curriculum Center (MCC) to serve school districts in the U.S. interested in mathematics curricula programs consistent with the Standards. The Center provides information and seminars for implementing 13 comprehensive mathematics programs funded by NSF. For further information on the programs listed below, contact MCC at (800) 332-2429 or through www.edc.org/mcc.

- ◆ Elementary School Projects
  - Everyday Mathematics (K–6)
  - Investigations in Number, Data, and Space (K–5)
  - Math Trailblazers (K–5)
- ◆ Middle School Projects
  - Connected Math (6–8)
  - Mathematics in Context (5–8)
  - MathScape: Seeing and Thinking Mathematically (6–8)
  - MathThematics: Six through Eight mathematics (STEM)
  - Middle School Mathematics Through Applications (MMAP) (6–8)
- ◆ Secondary School Projects
  - Contemporary Mathematics in Context (Core-Plus)
  - Interactive Mathematics Project (IMP)
  - Mathematics: Modeling Our World (ARISE)
  - MATH Connections
  - Connected Geometry
  - SIMMS Integrated Mathematics

REFERENCES

Ball, D. (1998). *The Subject Matter Preparation of Prospective Teachers: Challenging the Myth*. East Lansing, MI: National Council for Research in Teacher Education.

___. (1997). Developing Mathematics Reform: What Don't We Know about Teacher Learning—But Would Make Good Working Hypotheses. In *Reflecting on Our Work: NSF Teacher Enhancement in K–6 Mathematics*, edited by Susan N. Friel and George W. Bright. Lanham, MD: University Press of America.

Burrell, B., J. Miller, T. Pikes, and W. Sharpton. (1998). *Conversations with Author*. New Orleans, LA: University of New Orleans.

Burrill, G. (1998, April). *Conversations with Author*. NCTM Annual Conference. Washington, DC.

___. (1997). Choices and Challenges. *Teaching Children Mathematics, 4*(1), 58–63.

___. (1996, July/August). President's Message Column. *National Council of Teachers of Mathematics News Bulletin, 3*.

Caine, R.N., and G. Caine. (1997). *Education: On the Edge of Possibility*. Alexandria, VA: Association for Supervision and Curriculum Development.

Chapin, S. (1997). Introduction. In *The Partners in Change Handbook: A Professional Development Curriculum in Mathematics*. Boston, MA: Boston University Press.

Checkley, K., L. Kelly. (1999). Toward Better Teacher Education: A Conversation with Asa Hillard. *Educational Leadership, 8*, 58–62.

Cobb, P., T. Wood, and E. Yackel. (1990). Classrooms as Learning Environments for Teachers and Researchers. In *Constructivist Views on the Teaching and Learning of Mathematics*, edited by R.B. Davis, C.A. Maher, and N. Noddings. Reston, VA: National Council of Teachers of Mathematics, 125–146.

Coeyman M. (2000, May 23). U.S. School, Japanese Methods. *Christian Science Monitor*

Confrey, J. (1990). What Constructivism Implies for Teaching. In *Constructivist Views on the Teaching and Learning of Mathematics*, edited by R.B. Davis, C.A. Maher, and N. Noddings. Reston, VA: National Council of Teachers of Mathematics, 107–122.

Cooper, R. (1999, September 22). A Call to Focus on Techniques of Teaching. *Los Angeles Times*.

Cuevas, G. (1991). Developing Communication Skills in Mathematics for Students with Limited English Proficiency. *The Mathematics Teacher, 84*(3), 186–190.

Davidson, E., and J. Hammerman. (1993). Homogenized is Only Better for Milk. In *Reaching All Students With Mathematics*, edited by G. Cuevas and M. Driscoll. Reston, VA: National Council of Teachers of Mathematics, 197–211.

Fensham, P., R. Gunstone, and R. White. (1994). *The Content of Science: A Constructivist Approach to Its Teaching and Learning*. Washington, DC: Falmer Press.

Fuys, D., D. Geddes, and R. Tischler. (1988). The van Hiele Model of Thinking in Geometry Among Adolescents. Special issue of *The Journal for Research in Mathematics Education*, Monograph no. 3. Reston, VA: National Council of Teachers of Mathematics.

Goldin, G. A. (1990). Epistemology, Constructivism, and Discovery Learning Mathematics. In *Constructivist Views on the Teaching and Learning of Mathematics*, edited by R.B. Davis, C.A. Maher, and N. Noddings. Reston, VA: National Council of Teachers of Mathematics: 19–30.

Goleman, D. (1995). *Emotional Intelligence*. New York: Bantam Books.

Hoff, D. (2000, February 23). A Teaching Style That Adds Up. *Education Week*, 32–37. http://www.edweek.org/ew/ewstory.cfm?slug=24timss.h19.

Kaput, J.J. (1993). Technology and Mathematics Education. In *Handbook of Research on Mathematics Teaching and Learning*, edited by D. A. Grouws. New York: Macmillan, 515–555.

Khisty, L. L. (1997). Making Mathematics Accessible to Latino Students. In *NCTM 1997 Yearbook, Multicultural and Gender Equity in the Mathematics Classroom: The Gift of Diversity*. Reston, VA: National Council of Teachers of Mathematics, 92–101.

Lampert, M., and D. Ball. (1998). *Teaching, Multimedia, and Mathematics: Investigations of Real Practice*. New York: Teachers College Press.

Lappan, G. (1998, May/June). President's Message Column. *National Council of Teachers of Mathematics News Bulletin* 3.

Lenz, B. K., D.D. Deshler, and J. B. Schumaker. (1990). *The Development and Validation of Planning Routines to Enhance Delivery of Content to Students with Handicaps in General Education Settings. Progress Report.* Lawrence, KS: University of Kansas Center for Research on Learning.

Leonard, L. M., and M. D. Tracy. (1993). Using Games. *The Arithmetic Teacher,* 40(9), 499–503.

Lewis, C. (2000, April). Lesson Study: The Core of Japanese Development. Invited address to the Special Group on Research in Mathematics Education American Educational Research Association Meetings, New Orleans, session 47.09.

Linton Professional Development Corporation. (2000). Teaching Mathematics to Increase Student Achievement. *The Video Journal of Education,* 9(5).

Ma, L. (1999). *Knowing and Teaching Elementary Mathematics: Teachers' Understanding of Fundamental Mathematics in China and the U.S.* Hillsdale, NJ: Lawrence Erlbaum.

McGhan, B. (1995, October). Can It Be True That All Students Can Do Mathematics? *National Council of Supervisors of Mathematics Newsletter,* 9-13.

National Board for Professional Teaching Standards. (2000). *An Invitation to National Board Certification: What Every Teacher Should Know about the National Board Certification Process.* Southfield, MI: National Board for Professional Teaching Standards.

National Council of Teachers of Mathematics. (2000). *The Principles and Standards for School Mathematics.* Reston, VA: Author.

___. (1989). *Curriculum and Evaluation Standards for Learning Mathematics.* Reston, VA: Author.

___. (1991). *Professional Standards for Teaching Mathematics.* Reston, VA: Author.

___. (1995). *Assessment Standard for School Mathematics.* Reston, VA: Author.

National Science Teachers Association. (1977). *Games for the Science Classroom: An Annotated Bibliography.* Washington, DC: Author.

Nussbaum, J., and S. Novak. (1982). Alternative Frameworks, Conceptual Conflict and Accommodation: Toward a Principled Teaching Strategy. *Instructional Science, 11,* 183–200.

Owens, K., and R. Sanders. (1992). The Effectiveness of Games for Educational Purposes. A Review of Recent Research. *Simulation and Gaming*, 23(3), 261–276.

Phillip, R. and P. Schappelle. (1999). Algebra as Generalized Arithmetic: Starting with the Known for a Change. *Mathematics Teacher*, 92(4), 310–315.

Piaget, J. (1973). *To Understand is to Invent*. New York: Grossman.

Post, T., M. Behr, and R. Lesh, (1986). Research-Based Observations About Children's Learning of Rational Number Concepts. *Focus on Learning Problems in Mathematics*, 8(1).

Reeves, C. (2000). The Chicken Problem. *Mathematics Teaching in the Middle Schools*, 5(6), 398–402.

Sawada, D. (1997). NCTM's Standards in Japanese Elementary Schools. *Teaching Children Mathematics Teacher*, 4(1), 20–23.

Saul, M. (1997). Common Sense: The Most Important Standard. *The Mathematics Teacher*, 90(3), 182–184.

Schifter, D. (ed.) (1996a). *What's Happening in Math Class? Vol. 1. Envisioning New Practices Through Teacher Narratives*. New York: Teachers College Press.

___. (1996b). *What's Happening in Math Class? Vol. 2. Reconstructing Professional Identities*. New York: Teachers College Press.

Shapiro, L. A. (1992). *We're Number One*. New York: Vintage Books.

Schappelle, B., and R. Phillip. (1999). Algebra as Generalized Arithmetic: Starting from the Known for a Change. *The Mathematics Teacher*, 92(4), 310–316.

Shriner, J., T. Ysseldyke, and M. Honetschalger. (1994, March). "All" Means "All"—Including Students with Disabilities. *Educational Leadership*, 38–42.

Silver, E. (1997). "Algebra for All"—Increasing Students' Access to Algebraic Ideas, Not Just Algebra Courses. *Mathematics Teaching for the Middle Grades*, 2(4), 204–207.

Simon, A., R. Tzur, K. Heinz, M. Kinzel, and M. Schwan Smith. (2000). Characterizing a Perspective Underlying the Practice of Mathematics Teachers in Transition. *Journal for Research in Mathematics Education*, 31(5), 580–600.

Simon, M. A. (1995). Reconstructing Mathematics Pedagogy from a Constructivist Perspective. *Journal for Research in Mathematics Education*, 26(2), 114–145.

Slavin, R. E. (1990). *Cooperative Learning: Theory, Research, and Practice.* Englewood Cliffs, NJ: Prentice Hall.

Stevenson, H., and J. Stigler. (1992). *The Learning Gap: What our Schools Are Failing and What We Can Learn from Japanese and Chinese Education.* New York: Touchstone Books.

Stiff, L. V. (2000a, May/June). President's Message Column. *National Council of Teachers of Mathematics News Bulletin, 36,* 10.

___. (2000b, November). President's Message Column. *National Council of Teachers of Mathematics News Bulletin, 37,* 4.

___. (1993). Reaching All Students: A Vision of Learning Mathematics. In *Reaching All Students With Mathematics,* edited by G. Cuevas and M. Driscoll. Reston, VA: National Council of Teachers of Mathematics, 3–6.

Stigler, J. and J. Hiebert. (1999). Understanding and Improving Mathematics Instruction: An Overview of the TIMSS Video Study. *Phi Delta Kappan, 79(1),* 14–21.

___. (1999). *The Teaching Gap.* New York: Free Press

Stipek, D., M. J. Salmon, B. K. Givvin, E. Kazemi, G. Saxe, and L. V MacGyvers. (1998). The Value (and Convergence) of Practices Suggested by Motivation Research and Promoted by Mathematics Education References. *The Journal for Research in Mathematics Education, 29(4),* 465–488.

Strong, S. D., and B. N. Cobb. (2000, April). Algebra for All: It's a Matter of Equity, Expectations, and Effectiveness. *The Mathematics Education Dialogues, 3,* 2

Stuart, B. V. (2000). Math Curse or Math Anxiety? *Teaching Children Mathematics Teacher, 6(5),* 330–335.

Usiskin, Z. (1994, Winter). Individual Differences in the Teaching and Learning of Mathematics. *University of Chicago School Mathematics Improvement Newsletter,* 14.

Vygotsky, L. S. (1978). *Mind in Society: The Development of Higher Psychological Processes.* Cambridge, MA: Harvard University Press.

Wheately, G. H. (1991). Constructivist Perspectives on Science and Mathematics Learning. *Science Education, 75(1),* 9–21.

Williams, E. (1980). *An Investigation of Senior High School Students' Understanding of the Nature of Mathematical Proof.* Doctoral dissertation, University of Alberta.

Yoshida, M. (1999a). Lesson study: *An Ethnographic Investigation of School-Based Teacher Development in Japan.* Doctoral dissertation, University of Chicago.

___. (1999b). *Lesson Study in Elementary School Mathematics in Japan: A Case Study.* Paper presented at the American Educational Research Association Annual Meeting, Montreal, Canada.